EC&M

Understanding NE Code Rules On

EMERGENCY & STANDBY POWER SYSTEMS

Based on the 1996 National Electrical Code
Edited by Frederic P. Hartwell
Senior Editor, EC&M Magazine

Understanding NE Code Rules On Emergency and Standby Power Systems
Copyright ©1996 INTERTEC Publishing Corporation
All rights reserved

First Printing: April 1996

Published by
EC&M Books
INTERTEC Publishing Corp.
9800 Metcalf Avenue
Overland Park, KS 66212-2215

ISBN 0-87288-635-2

Please Note: The designations "National Electrical Code," "NE Code,"
and "NEC," where used in this book, refer to the National Electrical Code®,
which is a registered trademark of the National Fire Protection Association.

CONTENTS

Chapter 1.
ARTICLE 700 — EMERGENCY SYSTEMS 1
ARTICLE 700, PART A
Emergency Systems — General 1
  Sec. 700-1 — Scope 1
  Sec. 700-2 — Application of other articles 2
  Sec. 700-3 — Equipment approval 2
  Sec. 700-4 — Tests and maintenance 3
  Sec. 700-5 — Capacity 3
  Sec. 700-6 — Transfer Equipment 5
  Sec. 700-7 — Signals 7
  Sec. 700-8 — Signs 7
ARTICLE 700, PART B
Emergency Systems — Circuit Wiring 9
  Sec. 700-9 — Emergency system wiring 9
ARTICLE 700, PART C
Emergency Systems — Sources of Power 12
  Sec. 700-12 — General requirements 12
ARTICLE 700, PART D
Emergency Systems — Circuits for Lighting and Power 19
  Sec. 700-15 — Loads on emergency branch circuits 19
  Sec. 700-16 — Emergency illumination 19
  Sec. 700-17 — Circuits for emergency lighting 20
  Sec. 700-18 — Circuits for emergency power 21
ARTICLE 700, PART E
Emergency Power — Emergency Lighting Circuits 21
  Sec. 700-20 — Switch requirements 21
  Sec. 700-21 — Switch location 22
  Sec. 700-22 — Exterior lights 22
ARTICLE 700, PART F
Emergency Systems — Overcurrent Protection 22
  Sec. 700-25 — Accessibility 22
  Sec. 700-26 — Ground-fault protection of equipment 23

NFPA 110 – Standard for Emergency and Standby Power Systems (1993 edition) 23

Chapter 2.
ARTICLE 701 — LEGALLY REQUIRED STANDBY SYSTEMS 25
ARTICLE 701, PART A
Legally Required Standby Systems — General 25
    Sec. 701-1 — Scope 25
    Sec. 701-2 — Legally required standby systems 25
    Sec. 701-3 — Application of other articles 26
    Sec. 701-4 — Equipment approval 27
    Sec. 701-5 — Tests and maintenance 27
    Sec. 701-6 — Capacity and rating 28
    Sec. 701-7 — Transfer equipment 29
    Sec. 701-8 — Signals 31
    Sec. 701-9 — Signs 31
ARTICLE 701, PART B
Legally Required Standby Systems — Circuit Wiring 32
    Sec. 701-10 — Wiring legally required standby systems 32
ARTICLE 701, PART C
Legally Required Standby Systems — Sources of Power 32
    Sec. 701-11 — Legally required standby systems 32
ARTICLE 701, PART D
Legally Required Standby Systems — Overcurrent Protection 37
    Sec. 701-15 — Accessibility 37
    Sec. 701-17 — Ground-fault protection of equipment 37

Chapter 3.
ARTICLE 702 — OPTIONAL STANDBY SYSTEMS 39
ARTICLE 702, PART A
Optional Standby Systems — General 39
    Sec. 702-1 — Scope 39
    Sec. 702-2 — Optional standby systems 39
    Sec. 702-3 — Application of other articles 40
    Sec. 702-4 — Equipment approval 41
    Sec. 702-5 — Capacity and rating 41
    Sec. 702-6 — Transfer equipment 41

Sec. 702-7 — Signals 43
Sec. 702-8 — Signs 43
ARTICLE 702, PART B
Optional Standby Systems — Circuit Wiring 44
Sec. 702-9 — Wiring optional standby systems 44

Chapter 4.
DIFFERENCE BETWEEN EMERGENCY AND
STANDBY SYSTEMS 45
Legal requirements 45
Optional standby systems 48
Example 56

Chapter 5.
ARTICLE 705 — INTERCONNECTED ELECTRIC
POWER PRODUCTION SOURCES 57
Sec. 705-1 — Scope 57
Sec. 705-2 — Definition 58
Sec. 705-3 — Other articles 58
Sec. 705-10 — Directory 59
Sec. 705-12 — Point of connection 59
Sec. 705-14 — Output characteristics 61
Sec. 705-16 — Interrupting and withstand rating 62
Sec. 705-20 — Disconnecting means, sources 62
Sec. 705-21 — Disconnecting means, equipment 62
Sec. 705-22 — Disconnect device 63
Sec. 705-30 — Overcurrent protection 63
Sec. 705-32 — Ground-fault protection 67
Sec. 705-40 — Loss of primary source 69
Sec. 705-42 — Unbalanced interconnections 70
Sec. 705-43 — Synchronous generators 70
Sec. 705-50 — Grounding 70

Chapter 6.
ARTICLE 445 (GENERATORS) and ARTICLE 250
(GROUNDING) 73
Sec. 445-1 — General 74
Sec. 445-2 — Location 74

Sec. 445-3 — Marking 76
Sec. 445-4 — Overcurrent protection 76
Sec. 445-5 — Ampacity of conductors 78
Sec. 445-6 — Protection of live parts 80
Sec. 445-7 — Guards for attendants 80
Sec. 445-8 — Bushings 81
ARTICLE 250
Grounding 81
Sec. 250-5(d) — Separately derived systems 81
Sec. 250-6 — Portable and vehicle-mounted generators 83

Chapter 7.
ARTICLE 690 — SOLAR PHOTOVOLTAIC SYSTEMS 85
ARTICLE 690, PART A
Solar Photovoltaic Systems — General 85
Sec. 690-1 — Scope 85
Sec. 690-2 — Definitions 85
Sec. 690-3 — Other articles 87
Sec. 690-4 — Installation 88
Sec. 690-5 — Ground fault detection arrangement 89
ARTICLE 690, PART B
Solar Photovoltaic Systems — Circuit Requirements 89
Sec. 690-7 — Maximum voltage 89
Sec. 690-8 — Circuit sizing and current 90
Sec. 690-9 — Overcurrent protection 92
ARTICLE 690, PART C
Solar Photovoltaic Systems — Disconnecting Means 92
Sec. 690-13 — All conductors 92
Sec. 690-14 — Additional provisions 94
Sec. 690-15 — Disconnection of photovoltaic equipment 94
Sec. 690-16 — Fuses 94
Sec. 690-17 — Switch or circuit breaker 95
Sec. 690-18 — Disablement of an array 95
ARTICLE 690, PART D
Solar Photovoltaic Systems — Wiring Methods 95
Sec. 690-31 — Methods permitted 95
Sec. 690-32 — Component interconnections 97
Sec. 690-33 — Connectors 97

Sec. 690-34 — Access to boxes 98
ARTICLE 690, PART E
Solar Photovoltaic Sources — Grounding 98
Sec. 690-41 — System grounding 98
Sec. 690-42 — Point of system grounding connection 98
Sec. 690-43 — Equipment grounding 99
Sec. 690-45 — Size of equipment grounding conductor............ 99
Sec. 690-47 — Grounding electrode system 100
ARTICLE 690, PART F
Solar Photovoltaic Sources — Marking... 100
Sec. 690-51 — Modules 100
Sec. 690-52 — Photovoltaic power sources 101
ARTICLE 690, PART G
Solar Photovoltaic Sources — Connection to Other Sources 101
Sec. 690-61 — Loss of system voltage 101
Sec. 690-62 — Ampacity of neutral conductor 101
Sec. 690-63 — Unbalanced interconnections 102
Sec. 690-64 — Point of connection 103
ARTICLE 690, PART H
Solar Photovoltaic Systems — Storage Batteries 103
Sec. 690-71 — Installation 103
Sec. 690-72 — State of charge 105
Sec. 690-73 — Grounding 105
Sec. 690-74 — Battery interconnections 105

Chapter 8.
ARTICLE 517 — HEALTH CARE FACILITIES 107
ARTICLE 517, PART A
Health Care Facilities — General 107
Sec. 517-3 — Definitions 107
ARTICLE 517, PART B
Health Care Facilities — Wiring and Protection 109
Sec. 517-10 — Applicability 109
Sec. 517-17 — Ground-fault protection 110
Sec. 517-18 — General care areas 111
Sec. 517-19 — Critical care areas 111
ARTICLE 517, PART C
Essential Electrical System 111
Sec. 517-25 — Scope 111

Sec. 517-30 — Essential electrical systems for hospitals 112
Sec. 517-31 — Emergency system 117
Sec. 517-32 — Life safety branch 117
Sec. 517-33 — Critical branch 118
Sec. 517-34 — Equipment system connection to alternate
    power source 120
Sec. 517-35 — Sources of power 122
Sec. 517-40 — Essential electrical systems for nursing homes
    and limited care facilities 123
Sec. 517-41 — Essential electrical systems
    (nursing homes, etc.) 124
Sec. 517-42 — Automatic connection to life safety branch 125
Sec. 517-43 — Connection to critical branch 126
Sec. 517-44 — Sources of power 127
Sec. 517-50 — Systems for other health care facilities 128
NFPA Standards 129
NFPA 101 — Life Safety Code 132

Chapter 9.
ARTICLE 695 — FIRE PUMPS 135
Sec. 695-1 — Scope 135
Sec. 695-3 — Power source to motor driven fire pumps 135
Sec. 695-4 — Multiple power sources for firepump motors.... 137
Sec. 695-5 — Transformers 139
Sec. 695-7 — Equipment location 139
Sec. 695-8 — Power wiring 140
STANDARD NFPA 20 143
General 143
Diesel-engine drive 145
Engine-drive controllers 147
Acceptance testing, performance, and maintenance 150

PREFACE

A great number of compontents are involved in the installation of an auxiliary power system. The system selected and components involved depend on the type of occupancy, type of process or activity, and specific needs of the facility. Additional important considerations include the degree of safety or reliability desired, codes and standards, and the particular set of special functions and technologies required for the application.

The National Electrical Code provides guidance for safe and proper installation of systems required to provide emergency and legally required standby power sources. In addition, it also contains rules for those standby systems that are installed for the convenience of operations in a facility (optional standby systems), as well as those operated in parallel with an electric primary source and are capable of delivering energy to that source. Other NEC chapters also contain information that need to be understood in applying these onsite power sources.

But the NEC is not the only document that must be referred to when designing, installing, or operating standby power systems. The National Fire Protection Association (NFPA) have standards that apply, such as NFPA 99, 101, and 110. These are often referred to in fine print notes (FPNs) in applicable NEC chapters.

All these pertinent pieces of information have been gathered into this book, along with illustrations and examples of how the regulations are to be applied. The intent of this book is to simplify the task of those working on a project involving auxiliary standby power sources.

The following items are discussed in detail in this book.

Article 700 of the NEC covers systems that are legally required to be installed and that supply loads essential to safety and life.

Article 701 covers those systems required and classed as legally required standby by municipal, state, federal, or other codes or by any governmental agency having jurisdiction. These systems are intended to supply power automatically to important selected loads (other than those classed as emergency systems) in the event of failure of the normal source.

Article 702 deals with optional standby systems intended to protect private business or property where life safety does not depend on the performance of the system. They typically serve as an alternate power source for industrial and commercial buildings, farms, and residences.

Article 705 provides rules that apply to power sources that operate in parallel with the utility or other sources.

Article 517 includes numerous regulations that apply to onsite power systems installed in hospitals, nursing homes and other health-care facilities.

Article 690 gives the rules that apply to solar photovoltaic systems, which also are used to provide a source on onsite power.

Article 695 pulls together into one article the various regulations on fire pumps that were previously scattered throughout the NEC.

Article 455 provides general guidelines for proper and safe installation of generators.

This book contains a thorough explanation of the intent of each one of these articles. As mentioned previously, the most important rules of the NFPA standards are also discussed.

Overland Park, KS
April 1996

Alfred Berutti, P.E.
Editorial Projects Consultant, EC&M

ARTICLE 700 — EMERGENCY SYSTEMS

The first two parts of this article contain general rules that apply to the emergency system of all facilities. They are especially significant because the basic definition of what an emergency system is, when it is required, and how it is to be installed is defined there. In addition, Sec. 700-1 includes a reference to other documents that will be useful in understanding the subject. The applicable parts of these standards have been included in the this book.

ARTICLE 700, PART A
EMERGENCY SYSTEMS — GENERAL

Sec. 700-1 — Scope

Provisions of Article 700 apply to the electrical safety of the installation, operation, and maintenance of emergency systems. The circuits and equipment that are covered by these regulation are those that supply, distribute, and control electricity to facilities when the normal electrical supply or system is interrupted.

Here is the key statement made in this section. It clearly defines an "emergency system," thus differentiating it from a "legally required standby system" covered by Article 701 and an "optional standby system" covered by Article 702.

> Emergency systems are those systems legally required and classed as emergency by municipal, state, federal, or other codes, or by any governmental agency having jurisdiction. These systems are intended to automatically supply illumination or power, or both, to designated areas and equipment in the event of failure of the normal supply or in the event of accident to elements of a system intended to supply, distribute, and control power and illumination essential to safety to human life.

A fine print note (FPN No. 3) describes the following as places where emergency systems are applied.
- Places of assembly where artificial illumination is required

for safe exiting and for panic control in buildings subject to occupancy by large numbers of persons, such as hotels, theaters, sports arenas, health care facilities, and similar institutions.
- To provide power for such functions as ventilation where essential to maintain life, fire detection and alarm systems, elevators, fire pumps, public safety communication systems, industrial processes where power interruption would produce serious life safety or health hazards.
- Other similar functions.

Other FPNs following the definition refer to the following documents for further information on the subject of emergency systems.
- NEC Article 517 for emergency systems in health care facilities. This topic is covered in Chapter 9 of this book.
- NFPA 99 (*Standard for Health Care Facilities*) for emergency systems in those facilities.
- NFPA 101 (*Life Safety Code*) for where emergency lighting is essential.
- NFPA 110 (*Standard for Emergency and Standby Power Systems*) for performance requirements for the systems.

Fine print notes of the NEC are only advisory, not mandatory. Nonetheless, it's a good idea to be aware of the requirements shown in the listed documents. A condensed version of the most important points that deal with emergency systems is included in the appropriate chapters of this book.

Sec. 700-2 — Application of other articles

Except as modified by this article, all applicable articles of the NEC apply to emergency systems. In other words, this article is not the only one that applies to emergency systems. All provisions of Article 230 on services, Article 250 on grounding, etc. also apply except those sections that are specifically modified in Article 700.

Sec. 700-3 — Equipment approval

All equipment must be approved for use on emergency systems.

The term "approved" is defined in Article 100.

Approved: Acceptable to the authority having jurisdiction.

Use of listed or labeled equipment is most often the criteria used by the authority having jurisdiction for determining the suitability of equipment, although other benchmarks may be employed in deciding suitability.

Sec. 700-4 — Tests and maintenance

Unlike any other items covered by the Code, due to the critical nature of the equipment, the NEC lists specific tests and maintenance that must be carried out on emergency systems.

(a) Testing. The authority having jurisdiction is required by the Code to conduct or witness a test of the entire emergency system following completion of installation, and periodically afterwards.

(b) Frequency of testing. Emergency systems *must* be tested periodically on a schedule acceptable to the authority having jurisdiction to assure the systems are maintained in proper operating condition.

(c) Battery systems maintenance. Where battery systems or unit equipment are part of an emergency system, the authority having jurisdiction *must* require periodic maintenance. This includes batteries used for starting, control, or ignition of auxiliary engines, and those used in battery operated emergency lights and signs.

(d) Written record. A written record *must* be kept of tests and maintenance carried out on emergency systems.

(e) Testing under load. Means *must* be provided for testing all emergency lighting and power systems under maximum anticipated load conditions.

It is important to note that this schedule and prescription for carrying out the testing regime is mandatory. The actual provisions in the NEC use the word "shall," which signifies that the requirement "must" be carried out by the owner of the finished installation and the authority having jurisdiction.

Sec. 700-5 — Capacity

Specific rules are also included in Article 700 for assuring

Emergency power gensets do not need to be exclusively for emergency circuits; they can also be used for peak shaving and other purposes, but priority must be given to satisfying the emergency power requirements.

adequate capacity of emergency systems.

(a) Capacity and rating. An emergency system must have adequate capacity and rating for all loads that are to be operated simultaneously. The emergency system equipment must be suitable for the maximum available fault current at its terminals.

(b) Permitted uses. Alternate power sources are permitted to supply other than emergency system loads where automatic selective load pickup and load shedding is provided as needed to assure adequate power to the circuits in the following order of priority:

- emergency circuits;
- legally required standby circuits; and
- optional standby circuits.

Peak load shaving. Alternate power sources are also permitted to be used for peak load shaving, provided the same priority schedule is met.

In addition, operation of the alternate power source for peak load shaving is acceptable for satisfying the requirement for the frequency of testing spelled out in Sec. 700-4(b), provided all the other provisions of that section are met.

Availability. A portable or temporary alternate source must be available whenever the emergency generator is out of service for major maintenance or repair.

Sec. 700-6 — Transfer Equipment

Transfer equipment, including transfer switches, must be automatic and identified for emergency use and approved by the authority having jurisdiction. Transfer equipment must be designed and installed to prevent the inadvertent interconnection of normal and emergency sources of supply in any operation of the transfer equipment.

A reference is made to **Sec. 230-83**. There the requirement is that all ungrounded conductors of one source of supply must be disconnected before any ungrounded conductors of the second source is connected. This implies the use of an "open-transition" transfer switch that operates as shown in **Fig. 1.1**. Note that in both transfer and retransfer there is a stage when the load is connected to neither source, so there is no possibility of inadvertently interconnecting the normal and emergency sources.

Exceptions to this rule is made for cases:
- where manual equipment identified for the purpose, or suit-

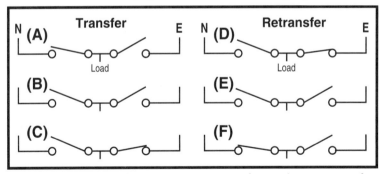

■ **Fig. 1.1.** Operation of an open-transition transfer switch. During transfer, in **(A)** the load is connected to the normal source (N); in **(B)** the load is disconnected from the normal source; and in **(C)** the load is connected to the emergency source (E). When normal power is reestablished, retransfer takes place. In **(D)** the load is still connected to the emergency power; in **(E)** the load is disconnected from the emergency source; and in **(F)** the load is connected to the normal source.

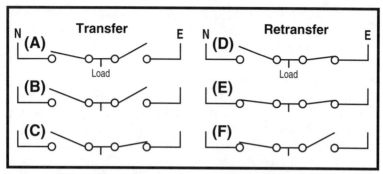

■ **Fig. 1.2.** Operating sequence of a closed-transition automatic transfer switch. When transfer is required because of failure of the normal source, there is no paralleling of sources. On retransferring, both sources are available and there is a brief period **(E)** during which the sources are paralleled.

able automatic equipment, is utilized; or
• where parallel operation is used and suitable automatic or manual control equipment is provided.

Permission to use other than open-transition transfer switches, or for two or more sources to be connected in parallel through the transfer equipment, allows the use of "closed-transition" transfer switches that connect both sources for a brief instant to prevent a

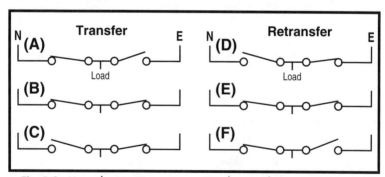

■ **Fig. 1.3.** Here, the emergency source is to be tested to assure it is capable of carrying the load. In the operating sequence, paralleling takes place both during the transfer sequence **(B)** and during retransfer **(E)**.

power interruption during transfer. This is shown in **Fig. 1.2** when the normal supply fails, and in **Fig. 1.3** when two active sources are available and the alternate source is being used for peak shaving, standby, or other operation, or the unit is to be tested in compliance emergency source requirements.

The permission also allows running an alternate source in parallel with a utility source provided adequate relaying is employed to assure the two are synchronized, etc. This allows for cogeneration between an emergency source and a utility source. This is discussed further in Chapter 5 of this book.

Sec. 700-7 — Signals

Audible and visual signaling devices must be provided (where practicable) to indicate the following:
- derangement of the emergency source;
- battery is carrying load;
- battery charger is not functioning; and
- ground fault on a solidly grounded wye emergency system of more than 150V to ground and circuit protective devices rated 1000A or more.

The sensor for the ground-fault signaling device must be located at, or ahead of, the main system disconnecting means for the emergency source. The maximum setting of the signaling device must not exceed a ground-fault current of 1200A.

Instructions on the course of action to be taken in event of a ground fault indication or alarm must be located at or near the sensor location.

An FPN refers to NFPA 110, for signals required for generator sets. These requirements are discussed later in this chapter.

Sec. 700-8 — Signs

Because of the critical nature of emergency systems, signs are required to be posted clearly identifying its various components.

(a) Emergency sources. A sign must be placed at the service entrance equipment identifying the type and location of onsite

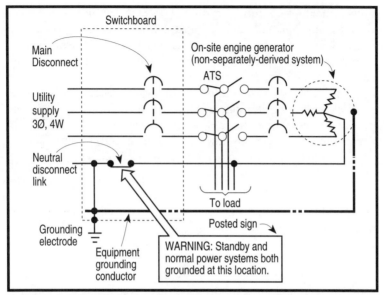

■ **Fig. 1.4.** When the connection to the grounding electrode conductor is remote from the emergency source, a sign must be posted at the connection warning that disconnection will affect the emergency source neutral integrity.

emergency power sources. An exception exempts unit equipment, such as an emergency light unit, from this requirement.

(b) Neutral grounding. If the grounded circuit conductor (neutral) from an emergency source is connected to a grounding electrode conductor anywhere except at the emergency source, as shown in **Fig. 1.4,** then a sign must be posted at the location of the remote grounding connection identifying all emergency and normal sources connected at that location.

This requirement is intended to prevent inadvertent interruption of the emergency source connection to the grounding electrode conductor while work is being done on the normal power supply. In the illustration, opening the disconnecting link interrupts the neutral grounding of both the normal power and emergency power.

ARTICLE 700, PART B

EMERGENCY SYSTEMS — CIRCUIT WIRING
Sec. 700-9 — Emergency system wiring

Wiring is critical to the operation of emergency systems. Thus, the NEC sets down some rules that will help make it obvious to anyone in construction, maintenance, or operations, that the installed circuits are part of an emergency system.

(a) Identification. All boxes and enclosures for emergency circuits must be permanently marked so they will be readily identified as a component of an emergency circuit or system. This identification requirement also applies to transfer switches, generators, power panels, etc.).

(b) Wiring. Emergency source wiring to emergency loads must be kept entirely independent of all other wiring and equipment and is prohibited from entering the same raceway, cable, box, or cabinet with other wiring.

There are several exceptions to this rule.

Exception No. 1. In transfer equipment enclosures.

Exception No. 2. In exit or emergency lighting fixtures supplied from two sources.

Exception No. 3. In a common junction box attached to exit or emergency lighting fixtures supplied from two sources.

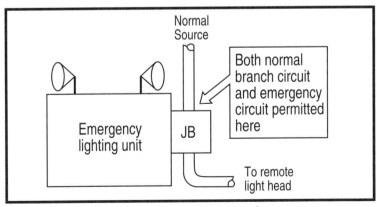

■ **Fig. 1.5.** While the general rule is that normal and emergency circuits must be kept separate, there as several permitted exceptions, including this one.

Exception No. 4. Wiring of two or more emergency circuits supplied from the same source.

Exception No. 5. As shown in **Fig. 1.5** on the previous page, in a common junction box attached to unit equipment that contains only the branch circuit supplying the unit equipment and the emergency circuit supplied by the unit equipment.

The exception (No. 1) that allows other wiring to be in the same transfer switch enclosure as emergency circuits is often misunderstood. Its purpose is to allow the normal source of power to be in the same transfer switch as the circuit from the emergency source,

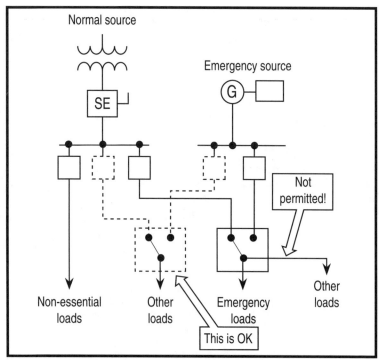

■ **Fig. 1.6.** Transfer switches in circuits feeding emergency loads must only contain wiring intended to supply these loads. Tapping within the enclosure to feed other nonemergency loads is prohibited.

which is obviously necessary in the transfer switch feeding the emergency loads. Note, however, that a sentence in that exception says that the emergency transfer switch must only feed emergency loads. This means that if other non-emergency loads are to be supplied by the emergency source, they must *not* be tapped off within the transfer switch as shown in **Fig. 1.6**. Rather, the non-emergency loads must be fed from their own transfer switch as shown.

Emergency wiring circuits must be designed and located to minimize the hazards that might cause failure due to flooding, fire, icing, vandalism, and other adverse conditions.

(c) Fire protection. Enhanced protection of emergency systems in higher hazard occupancies is required. Among the facilities covered by this provision are:

- occupancies for assembly of 1000 persons and over; and
- high-rise buildings over 75-ft (23 m) high that include places of assembly, educational, residential, business, mercantile, or detention and correctional facilities.

There are two ways in which this enhanced protection can be achieved. The first is by the installation of a fire suppression system that fully protects the emergency system feeder-circuit wiring and also the equipment such as panelboards, transfer switches, transformers, etc. Note, however, that just because a building is fully sprinklered in accordance with the local building codes does not necessarily mean that this rule is met. If the feeder rises through a chase, then that chase must also be protected by the automatic fire suppression system.

The second way to protect the emergency system is shown in **Fig. 1.7** on the next page. Here the emergency source and transfer switch are within a space with a minimum 1-hr fire rating, and the feeder is run in an *electrical circuit protective system* having a 1-hr fire rating. These systems are listed in the UL *Building Material Directory*. They include a special wrap for 2-in. or larger steel rigid metal conduit, and type MI cable with special additional provisions such as being supported every 3 ft and with all terminating seals being at least 1 ft within a protected area.

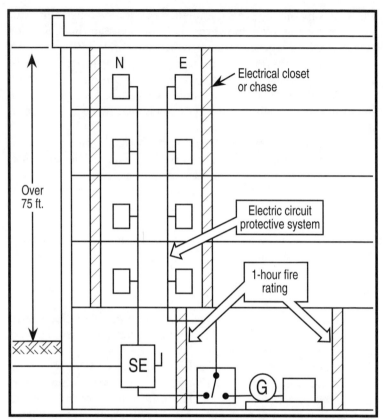

■ **Fig. 1.7.** Enhanced protection is required for feeders to emergency equipment. One of the options is shown here for cases where a fire suppression system is not employed. Depending upon the actual configuration of the emergency system, a combination of fire suppression system, fire-resistance ratings, and electrical circuit protective systems may have to be used.

ARTICLE 700, PART C
EMERGENCY SYSTEMS — SOURCES OF POWER

Sec. 700-12 — General requirements

The power supply must be such that, in the event of failure of the normal supply to or within a building or group of buildings where emergency power is deemed necessary, emergency lighting, emergency power, or both, will be available within the time required for the application, but not to exceed 10 seconds.

In selecting an emergency source of power, consideration must be given to the occupancy and the type of service to be rendered. The decision to be made is whether the emergency power is only required to be of minimum duration, as would be the case for a theater. There, the emergency power is only required to be maintained for a long-enough period for the facility to be evacuated. On the other hand, emergency power may be required for a longer period, such as for supplying power and lighting during an indefinite period of primary power interruption caused by trouble either inside or outside a building.

An FPN states that assigning the degree of reliability required of the recognized emergency supply system depends upon a careful evaluation of the variables at each particular installation. In other words, there are no set rules that can cover the many possible situations. A thorough analysis of each individual need must be made.

Equipment must be designed and located to minimize the hazards that might cause complete failure due to, among other things:
- flooding;
- fires;
- icing; and
- vandalism.

Language similar to that of Sec. 700-9(c) is included in this section as a guide on how the emergency generating equipment can be protected against fire. It can be either guarded by a fire suppression system, or be within a space having a minimum 1-hr fire rating.

One or more of the types of power supply systems that meet the following requirements is permitted to be used as part of an emergency power system.

(a) Storage batteries used as a source of power for emergency systems must have a suitable rating and capacity to supply and maintain the total load for a period of $1^1/_2$ hours minimum, without the voltage applied to the load falling below $87^1/_2\%$ of normal.

Batteries, whether of the acid or alkali type, must be designed

Large battery banks are used as a source for supplying emergency AC power via an inverter, or to supply DC directly to emergency tripping circuits and other critical loads.

and constructed to meet the requirements of emergency service and be compatible with the charger for that particular installation. An automatic battery charging means must be provided.

A lead-acid battery that requires water additions must be furnished with transparent or translucent jars. Automotive-type batteries must not be used. For a sealed battery, the container is not required to be transparent.

(b) Generator set. A generator set driven by a prime mover acceptable to the authority having jurisdiction, and having a capacity and rating adequate for all emergency loads that are to be operated simultaneously, must also meet the following conditions.

Starting. Means must be provided for automatically starting the prime mover on failure of the normal service and for its automatic transfer onto the line to pick up all required electrical circuit loads.

Retransfer. Once the genset is running and on line, a time-delay feature must prevent the emergency load from retransferring to the main power source for 15 minutes. This avoids retransfer on short-time reestablishment of the normal source via automatic circuit reclosers and similar non-permanent return of prime-source power.

Fuel supply. Where internal combustion engines are used as the prime mover, an onsite fuel supply must be provided having sufficient capacity for not less than 2-hrs of full-demand operation of the system.

Prime movers must not be solely dependent upon a public utility gas system for their fuel supply or municipal water supply for their cooling systems. Where dual fuel supplies are used, means must be provided for automatically transferring from one fuel supply to another.

An exception to the onsite fuel requirement say that if acceptable to the authority having jurisdiction, the use of other than onsite fuels is permitted where there is a low probability of a simultaneous failure of both the off-site fuel delivery system and power from the outside electrical utility company.

Batteries. Where a storage battery (or a string of batteries) is used for control or signal power, or as the means of starting the prime mover, it must be suitable for the purpose. Remember that according to the definition, a "listed" product either meets appropriate standards, or has been tested and found suitable for use in a specified manner. The battery must be equipped with an automatic charging means independent of the generator set.

Availability. Generator sets that require more than 10 seconds to develop power are acceptable, provided an auxiliary power supply will energize the emergency system until the generator can pick up the load. This provision most often applies to large installations where a UPS system will instantaneously pick up the load and carry it for a fixed period of time. The genset is only used to supply power for outages lasting longer than the rating of the UPS system, or upon its failure to perform properly. Starting time, thus, is not critical.

(c) Uninterruptible power supply. UPS systems used to provide power for emergency systems must comply with the applicable provisions of Sec. 700-12(a) with regard to the batteries that are part of the system, and Sec. 700-12(b) on how an engine-generator set is to be used as a backup source.

Uninterruptible power supplies (UPS) are among the types of

equipment that are also recognized by Sec. 701-11(c) as a source of legally required standby power. There are few specific rules in the NEC that govern the application of a UPS system.

When located within a "computer room," however, **Sec. 645-11** must be complied with. The requirement given in that section is that UPS systems installed within the electronic computer/data processing room, and their supply and output circuits, must comply with **Sec. 645-10**, which deals with disconnecting means.

In Sec. 645-10, it says that a means must be provided to disconnect power to all electronic equipment in the electronic computer/data processing equipment room. It must also disconnect the battery from its load.

Controls of these disconnecting means are to be grouped, identified, and be readily accessible at the principal exit doors. It is permitted to have this control combined with that for disconnecting the HVAC system of the room.

An exception exempts systems that qualify as being integrated electrical systems (per **Article 685**) from having a disconnect. These systems require an orderly shutdown process and, thus, would be adversely affected by a sudden cutoff of power.

UPS systems can provide emergency power to maintain power to computer-controlled process instrumentation and interface equipment. An exception exempts these facilities from the requirement of having a disconnect at the exit doors of a main control room if a sudden loss of power cannot be tolerated.

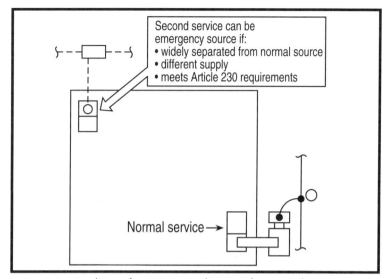

■ **Fig. 1.8.** Conditions for permitting the use of a second service as the emergency power source.

(d) Separate service. Where acceptable to the authority having jurisdiction as being suitable for use as an emergency source, a second service is permitted for this purpose. As shown in **Fig. 1.8**, this emergency service must be:

- in accordance with Article 230;
- from a separate service drop or lateral than the normal service; and
- widely separated electrically and physically from the normal service to minimize the possibility of simultaneous interruption of the supplies.

Sec. 230-2 specifies when multiple services are permitted. The general rule is that a building or other structure must be supplied by only one service. There are, however, many exceptions to this rule, including for a fire pump when required, and for emergency, legally required standby, optional standby, or parallel power production systems where a separate service is required.

When more than one service is permitted by any of these exceptions, a permanent plaque or directory must be installed at each service drop or lateral, or at each service-equipment location, identifying all the other services on or in the building or structure and describing the area served by each.

(e) Unit equipment in accordance with this subsection must satisfy the applicable requirements of this article.

Individual emergency lighting units (ELUs) for emergency illumination must consist of:
- a rechargeable battery;
- a battery charging means;
- provisions for one or more lamps mounted on the equipment, or terminals for remote lamps, or both; and
- a relaying device arranged to energize the lamps automatically upon failure of the supply to the unit equipment.

The batteries used in the ELU, whether acid or alkali type, must be designed and constructed to meet the requirements of emergency service. They must be of suitable rating and capacity to supply and maintain at not less than:
- $87^1/_2$% of the nominal battery voltage for the total lamp load associated with the unit for a period of at least $1^1/_2$ hours; or
- 60% of the initial emergency illumination for a period of at least $1^1/_2$ hours.

ELUs must be permanently fixed in place (not portable) and have all wiring to each unit installed in accordance with the requirements of any of the wiring methods listed in NEC Chapter 3. A flexible cord-and-plug connection is permitted, provided that the cord length does not exceed 3 ft (914 mm).

The ELU must be connected to the same branch circuit as that serving the normal lighting in the area. If this branch circuit contains local switching of the lighting fixtures in the area, the ELU must be connected ahead of the switches so that only an interruption of power at the panelboard will cause the ELU to be turned on.

An exception to this requirement (**Fig. 1.9**) is made for a separate and uninterrupted area supplied by a minimum of three normal

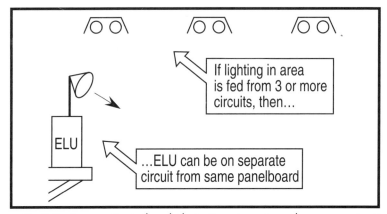

- **Fig. 1.9.** An exception to the rule that unit equipment must be on same circuit as lighting in the area.

lighting circuits. In such a case, a separate branch circuit for the ELUs in that area is permitted. But, the circuit for the ELU must originate from the same panelboard as that of the normal lighting circuits and its circuit breaker must be provided with a lock-on feature.

Branch circuits feeding ELUs must be clearly identified at the distribution panelboard.

Emergency illumination heads that obtain power from an ELU, but are mounted remote from the unit, must be wired to the ELU as described previously in Sec. 700-9, and by one of the wiring methods listed in NEC Chapter 3.

ARTICLE 700, PART D
EMERGENCY SYSTEMS — CIRCUITS FOR LIGHTING AND POWER

Sec. 700-15 — Loads on emergency branch circuits

No appliances and no lamps, other than those specified as required for emergency use, are permitted to be supplied by emergency lighting circuits.

Sec. 700-16 — Emergency illumination

Emergency illumination includes all required means of egress lighting, illuminated exit signs, and all other lights specified as necessary to provide required illumination. This statement really correlates with Sec. 700-15 because it defines what fixtures *are permitted* to be supplied by emergency lighting circuits.

Emergency lighting systems must be so designed and installed that the failure of any individual lighting element, such as the burning out of a lamp, cannot leave in total darkness any space that requires emergency illumination.

Where high-intensity discharge lighting (such as high- and low-pressure sodium, mercury vapor, and metal halide) is used as the sole source of normal illumination, the emergency lighting system is required to operate until normal illumination has been restored.

During the "restrike" period, HID lamps reach their full light output over a period of time, which can extend for a few minutes. For that reason, the emergency lighting system is required to be maintained in operation during that period. An exception to this rule is made for cases where alternative means have been taken to ensure that the emergency lighting illumination level is maintained. This could be achieved by installing a few strategically placed incandescent fixtures as part of the lighting pattern in an area, by HID fixtures containing a second small incandescent source that operates during the restrike period, or other means.

Sec. 700-17 — Circuits for emergency lighting

Branch circuits that supply emergency lighting are to be installed to provide service from a source complying with Sec. 700-12 (discussed previously in this chapter) when the normal supply for lighting is interrupted. Such installations must provide either one of the following:

• an emergency lighting supply, independent of the general lighting supply, with provisions for automatically transferring to the emergency lights upon failure of the general lighting system supply; or

• two (or more) separate and complete systems each with an independent power supply that provides sufficient current for emergency lighting purposes. Upon failure of one system, means must be provided for automatically energizing the other system — unless both systems are used for regular lighting purposes and are both kept lighted.

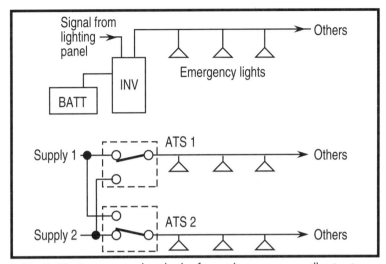

■ **Fig. 1.10.** Two permitted methods of providing emergency illumination.

Either or both systems illustrated in **Fig. 1.10** are permitted to be part of the general lighting system of the protected occupancy if circuits supplying light for emergency illumination are installed in accordance with other sections of this article.

Sec. 700-18 — Circuits for emergency power

For branch circuits that supply equipment classed as emergency, there must be an emergency supply source to which the load will be transferred automatically upon the failure of the normal supply.

As discussed earlier in this chapter, this source can be a generator that can be on line within 10 seconds of the power failure, storage batteries, a UPS, or a separate service.

ARTICLE 700, PART E
EMERGENCY POWER — EMERGENCY LIGHTING CIRCUITS

Sec. 700-20 — Switch requirements

The switch or switches installed in emergency lighting circuits must be so arranged that only authorized persons will have control of emergency lighting.

There are exceptions to this rule.

Exception No. 1. Where two or more single-throw switches are connected in parallel to control a single circuit, at least one of these switches must be accessible only to authorized persons.

Exception No. 2. Additional switches that act only to put emergency lights into operation, but not deenergize them, are permitted.

Sec. 700-21 — Switch location

All manual switches for controlling emergency circuits must be in locations convenient to authorized persons responsible for their actuation. In places of assembly, such as theaters, a switch for controlling emergency lighting systems must be located in the lobby or at a place conveniently accessible thereto.

In no case can a control switch for emergency lighting in a theater, motion-picture theater, or place of assembly be placed in a projection booth, on a stage, or a platform. An exception to this is that when multiple switches are provided, one can be placed in these locations if it can only energize the circuit, but not deenergize the circuit.

Sec. 700-22 — Exterior lights

Those lights on the exterior of a building that are not required for illumination when there is sufficient daylight are permitted to be controlled by an automatic light-actuated device.

ARTICLE 700, PART F
EMERGENCY SYSTEMS — OVERCURRENT PROTECTION
Sec. 700-25 — Accessibility

The branch-circuit overcurrent devices in emergency circuits must be accessible to authorized persons only.

An FPN explains that fuses and circuit breakers for emergency circuit overcurrent protection, where coordinated to ensure selective clearing of fault currents, increase overall reliability of the system.

Sec. 700-26 — Ground-fault protection of equipment

An alternate source for emergency systems is not required to be automatically disconnected by its ground-fault protection equipment. Instead, ground-fault indication of the emergency source is to be provided per Sec. 700-7(d), which has been covered earlier in this chapter.

NFPA 110
STANDARD FOR EMERGENCY AND STANDBY POWER SYSTEMS (1993 edition)

This standard provides specific rules for signals required with generator sets. They vary according to the "level" of criticality of the emergency power supply system. NFPA 110 recognizes that the systems are used in many different purposes and that the requirement for one application may not be appropriate in another. Thus, three levels are recognized.

Level 1 defines equipment performance where failure of equipment to perform could result in serious injuries or loss of human life. This type of equipment must be permanently installed.

Appendix A of NFPA 110 has a further definition of Level 1 that is similar to that contained in **Sec. 700-1** of the NEC in defining an *emergency system*.

Level 2 defines equipment performance where failure of the supply to perform is less critical to human life and safety and where it is expected that the authority having jurisdiction will exercise this option to allow a higher degree of flexibility. This equipment must also be permanently installed.

A further discussion in Appendix A of NFPA 110 uses essentially the same language as that in **Sec. 701-1** to define a *legally required standby system*.

Level 3 defines all other equipment and applications not defined as Level 1 or Level 2. This includes optional standby systems. NFPA 110 contains no requirements for this equipment.

NFPA 110, **Table 3-5.5.2(d)** lists the required (X) and optional (O) safety indications and shutdowns required for a Level 1 and Level 2 system. The needed indications are shown in **Fig. 1.11**.

	Level 1			Level 2		
Indicator Function (at Battery Voltage)	Control Panel	Remote Alarm	Shut Down	Control Panel	Remote Alarm	Shut Down
Overcrank	x	X	X	X	X	O
Low Water Temp. < 70°F (21°C)	X		X	X		O
High Engine Temperature Prealarm	X		X	O		
High Engine Temperature	X	X	X	X	X	O
Low Lube Oil Pressure Prealarm	X		X	O		
Low Lube Oil Pressure	X	X	X	X	X	O
Overspeed	X	X	X	X	X	O
Low Fuel Main Tank	X		X	O		O
EPS Supplying Load	X			O		
Control Switch not in Auto. Position	X		X	O		
Battery Charger Malfunctioning	X			O		
Low Voltage in Battery	X			O		
Lamp Test	X			X		
Contacts for Local and Remote Common Alarm	X		X	X		X
Audible Alarm Silencing Switch			X			O
Low Starting Air Pressure	X			O		
Low Starting Hydraulic Pressure	X			O		
Air Shutdown Damper when used	X	X	X	X	X	O
Remote Emergency Stop		X			X	

■ **Fig. 1.11.** Safety indication and shutdown requirements of NFPA 110 for generator sets applied in a Level 1 and a Level 2 system [adapted from Table 3-5.5.2(d)].

There are many other electrical requirements included in NFPA 110.
- Chapter 3 details required control functions;
- Chapter 4 gives electrical requirements;
- Chapter 5 deals with installation and environmental considerations;
- Chapter 6 covers routine maintenance and operational testing; and
- Appendix A further elaborates on requirements.

ARTICLE 701 — LEGALLY REQUIRED STANDBY SYSTEMS

2.

In discussing alternate power sources, it is important to remember the distinction between those that are required by statutes and those that are installed to meet the needs of a particular process or installation, but are not mandated by any law.

ARTICLE 701, PART A
LEGALLY REQUIRED STANDBY SYSTEMS — GENERAL
Sec. 701-1 — Scope

Provisions of this article apply to the electrical safety of circuits and equipment intended to supply, distribute, and control electricity within facilities that are legally required to provide illumination and/or power when the normal electrical supply is interrupted. The systems covered by this article include only those that are *permanently installed* in their entirely, including the power source.

The following references for additional information are given in FPNs:
- NFPA 99, *Health Care Facilities*;
- NFPA 110, *Emergency and Standby Power Systems*; and
- ANSI/IEEE 446, *Recommended Practice for Emergency and Standby Power Systems for Industrial and Commercial Applications.*

The main points in NFPA 99 are discussed in Chapter 9 of this book, and those of NFPA 110 that deal with legally required standby systems were discussed in Chapter 1 in this book.

ANSI/IEEE 446 also is known as the IEEE "Orange Book" that serves as a source book for both emergency and standby sources of power. This document is much too broad in scope to be summarized here. Where necessary, refer to the full text of that publication.

Sec. 701-2 — Legally required standby systems

Systems that meet this definition are those that are so classified by municipal, state, federal, or other codes, or by any governmental agency having jurisdiction. These systems are intended to automatically supply power to selected loads (other than those classed as emergency systems) in the event of failure of the normal source.

An FPN explains that legally required standby systems are typically installed to serve loads such as heating and refrigeration, communications, ventilation and smoke removal, sewerage disposal, lighting, and industrial processes. The criteria used in assessing that they belong in this classification is that when the system is stopped due to an interruption of the normal electrical supply:
- a hazard could be created; or
- the rescue or fire-fighting operations could be hampered.

Sec. 701-3 — Application of other articles

Except as modified by this article, all applicable articles of the

Industrial processes sometimes contain reactor vessels in which loss of electric power for an extended period would allow a dangerous runaway exothermic reaction to take place. In such a case, the vessel and its associated cooling water pumps, controls, and other associated equipment would be required to have a legally required standby system to back up the normal electrical supply.

NEC apply to legally required standby systems.

Articles such as 230 on services, 250 on grounding, Article 445 on generators, and others, must be complied with in installing, maintaining, and operating a facility containing a legally required standby system. Where any of the provisions of these other articles conflict with the requirements of Article 701, then the requirements of this article supersede those in other sections of the NEC.

Sec. 701-4 — Equipment approval

All equipment used in legally required standby systems must be approved for the intended use. Remember that the term "approved" means that the item is acceptable to the authority having jurisdiction. Thus, it is not permitted to include a power source or other items used as part of a legally required standby system without receiving the approval of the authority having jurisdiction. The purpose is to assure that, because of the critical nature of the equipment, the items meet the appropriate standards for this category of standby system.

Sec. 701-5 — Tests and maintenance

As with emergency systems, it is mandated that a legally required standby system be thoroughly tested before it is placed into service to provide a backup source of electrical power to selected loads. The following rules *must* be adhered to.

• Following its installation, the authority having jurisdiction is required to either conduct or witness the test on the *complete* system.

• Thereafter, the legally required standby system is to be tested periodically on a schedule and in a manner acceptable to the authority having jurisdiction to assure it has been maintained in proper operation condition.

• The authority having jurisdiction must also require periodic maintenance of batteries used for control, starting, or ignition of prime movers.

• Means for testing the legally required standby system *under load* is to be provided, and a written record must be kept of all tests

Proper maintenance of equipment that is part of a legally required standby system is mandated to assure that the system will operate properly when called on following a electrical power outage.

and maintenance performed to comply with the rules of Sec. 701-5.

Note that there is a difference between Sec. 700-4 and Sec. 701-5 in the wording of the need for participation by the authority having jurisdiction in the testing procedure. For emergency systems, the authority must witness or carry out the tests, "upon installation and periodically thereafter." For legally required standby systems, the authority only has to witness or conduct the test following the completion of installation, not thereafter.

Sec. 701-6 — Capacity and rating

A legally required standby system must have adequate capacity and rating for the supply of all equipment intended to be operated at one time. The system equipment must be suitable for the maximum available fault current at its terminals.

The alternate power source supplying the legally required standby system loads is also permitted to supply optional standby system loads. There is, however, a restriction to this permission. As shown

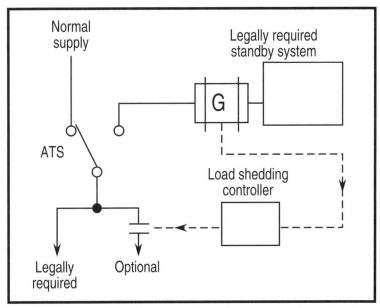

■ **Fig. 2.1.** The load-shedding controller will drop optional standby loads according to a preset priority list to keep the legally required standby system from being overloaded.

in **Fig. 2.1**, automatic selective load pickup and load shedding is to be provided as needed to assure adequate power to the legally required standby circuits.

In other words, the legally required standby system loads must be given first priority. If the alternate power source is incapable of fully carrying both the legally required and optional standby loads, then enough optional standby system loads must be automatically removed from the line to assure adequate power for the legally required system loads.

Sec. 701-7 — Transfer equipment

Automatic transfer switches, and the like, must be identified for standby use and be approved by the authority having jurisdiction. Transfer equipment must be capable of, and be installed in a manner, preventing the inadvertent interconnection of normal and alternate sources of supply in any operation of the transfer equipment.

A reference is made to **Sec. 230-83**, which states that transfer equipment, including transfer switches must operate in a manner that all ungrounded conductors of one source of supply are to be disconnected before any ungrounded conductors of the second source are connected. As with emergency systems, the exceptions to this rule also apply here.

Exception No. 1: Where manual equipment identified for the

An exception to the general rule (that transfer equipment prevent interconnection of normal and alternate sources) permits such operation of equipment that is approved for the purpose, such as this 1600A closed transition automatic transfer switch.

purpose or suitable automatic equipment is utilized, two or more sources are permitted to be connected in parallel through transfer equipment.

Exception No. 2: Where parallel operation is used and suitable automatic or manual control equipment is provided.

Means to bypass and isolate the transfer switch equipment is permitted. Where these bypass isolation switches are used, inadvertent parallel operation must be avoided.

Sec. 701-8 — Signals

Audible and visual signal devices must be provided (where practicable) to indicate:
- failure of malfunction of the standby source;
- that the standby source is carrying load; and
- that the battery charger is not functioning.

An FPN refers to NFPA 110 for the required signals for generator sets used in legally required standby service. The provisions were discussed in Chapter 1 of this book.

Sec. 701-9 — Signs

The following signs must be posted to assure that a legally-required standby system is not interrupted.
- A sign must be placed at the service entrance indicating the type and location of onsite legally required standby power sources.
- Where the grounded circuit conductor connected to the emergency source is connected to a grounding electrode conductor at a location remote from the emergency source, there must be a sign at the grounding location that identifies all emergency and normal sources connected at that location.

Both of these requirements duplicate the requirements for signs for an emergency system. Refer to the discussion under Sec. 700-8 in Chapter 1 of this book. Similar to emergency systems, unit equipment is exempted from the sign requirement.

ARTICLE 701, PART B
LEGALLY REQUIRED STANDBY SYSTEMS — CIRCUIT WIRING

Sec. 701-10 — Wiring legally required standby systems

Wiring of legally required standby systems is permitted to occupy the same raceways, boxes, and cabinets with other general wiring. This is in sharp contrast to wiring of emergency systems, where such an arrangement is prohibited.

ARTICLE 701, PART C
LEGALLY REQUIRED STANDBY SYSTEMS — SOURCES OF POWER

Sec. 701-11 — Legally required standby systems

In the event of failure of the normal supply to or within a building or group of buildings where legally required standby power is mandated, the alternate power supply must be available within the time required for the application, but not to exceed 60 seconds.

In NFPA 110, paragraph 2-3.2 defines a generator for this service as a Type 60 emergency power supply system. In selecting a legally required standby source of power, consideration must be given to the type of service to be rendered, whether of short-time or long duration. For generator type equipment, the minimum time in hours the system is designed to operate at its rated load without being refueled is assigned a "Class" designation in paragraph 2-3.3. These can vary from Class 0.083 (5 min) to Class 48 (48 hrs), and beyond to Class X (hours required by application).

The NEC says consideration also must be given to the location or design, or both, of all equipment to minimize the hazards that might cause complete failure due to floods, fires, icing, and vandalism.

An FPN states that assignment of the degree of reliability of a recognized legally required standby supply system depends upon the careful evaluation of the variables at each particular installation.

The supply system for legally required standby purposes, in addition to the normal services to the building, is permitted to comprise one of more of the following types of systems.

(a) Storage battery used as a source of power for legally required standby systems must be of suitable rating and capacity to

Battery reliability is essential to assure engine-generator sets can be started when needed. Shown is an *bad* installation that virtually assures *failure*. The batteries are not sealed types, yet are not in translucent jars; the battery rack and floor show signs of extensive acid spills; and the cables are inadequately supported.

supply and maintain the total load for a period of $1^{1}/_{2}$ hours minimum, without the voltage applied to the load falling below $87^{1}/_{2}\%$ of normal.

Batteries, whether of the acid or alkali type, must meet the requirements of emergency service and be compatible with the charger for that particular installation. An automatic battery charging means must be provided.

A lead acid battery that requires water additions must be furnished with transparent or translucent jars. Automotive-type batteries must not be used. For a sealed battery, the container is not required to be transparent.

(b) Generator set. A generator set driven by a prime mover acceptable to the authority having jurisdiction and having a capacity and rating adequate for all legally required standby system loads that are to be operated simultaneously must meet the following conditions.

- Means must be provided for automatically starting the prime mover on failure of the normal service and for its automatic transfer onto the line to pick up all required electrical circuit loads.

- Once the gen-set is running and on line, a time-delay feature must prevent the load from retransferring to the main power source for 15-min. This avoids retransfer on short-time reestablishment of the normal source via automatic circuit reclosers and similar non-permanent return of prime-source power.

- Where internal combustion engines are used as the prime mover, an *onsite* fuel supply must be provided having sufficient capacity for not less than 2 hours full-demand operation of the system (Class 2 per NFPA 110).

- Prime movers must not be solely dependent upon a public utility gas system for their fuel supply or municipal water supply for their cooling systems. Means must be provided for automatically transferring from one fuel supply to another where dual fuel supplies are used. This implies a requirement for an *onsite* gas and water supply.

An exception to the onsite fuel requirement say that, if acceptable to the authority having jurisdiction, the use of other than onsite fuels is permitted where there is a low probability of a simultaneous failure of both the off-site fuel delivery system and power from the outside electrical utility company.

- Where a storage battery is used for control or signal power, or as the means of starting the prime mover, it must be suitable for the purpose and must be equipped with an automatic charging means independent of the generator set.

(c) Uninterruptible power supply. UPS systems used to provide power for emergency systems must comply with the applicable provisions of Secs. 701-11(a) and (b).

The first of these requirements apply to batteries and their charger, which are an integral part of all UPS systems. The second applies principally to rotary UPS systems where a motor-generator set provides the required AC power. A static UPS system supplies the AC power via an inverter system. No specifics on these sources are included in Article 701 of the NEC or in NFPA 110. NFPA 110A includes details on inverters, but this document is not referenced in this article of the NEC.

For more information on static and rotary UPS systems see *EC&M's* book entitled, *Practical Guide to Quality Power*. For more information on UPS systems in computer rooms, see *EC&M's* book entitled, *Practical Guide to Power Distribution Systems for Computers*.

(d) Separate service. Where acceptable to the authority having jurisdiction, a second service is permitted for this purpose. This alternate service must be:

- in accordance with Article 230;
- from a separate service drop or lateral than the normal service; and
- widely separated electrically and physically from the normal service to minimize the possibility of simultaneous interruption of the supplies.

The same limitations on separate services apply as those discussed for emergency systems.

(e) Connection ahead of service disconnecting means. Where acceptable to the authority having jurisdiction as suitable for use as an alternate source, connection ahead of, but not within, the main service disconnecting means is permitted. The legally required standby service must be sufficiently separated from the normal main service disconnecting means to prevent simultaneous interruption of the supply due to something that occurs within the building or groups of buildings served.

Note that taps ahead of the main disconnect are no longer permitted to serve as the source of supply for emergency systems. It, however, continues to be permitted in legally required standby systems.

An FPN refers to **Sec. 230-82** for equipment permitted to be

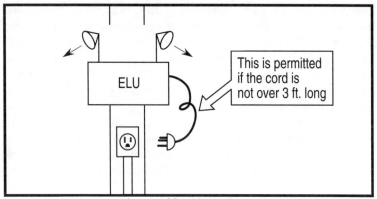

■ **Fig. 2.2.** Limitation on the use of flexible cord to connect an ELU to its power source.

connected on the supply-side of a service disconnect. The general rule there prohibits connections on the line-side of a service, but there are many exceptions to this rule. The one that applies here is:

Exception No. 5. Taps used only to supply ... standby power systems, ... if provided with service equipment and installed in accordance with requirements for service-entrance conductors.

(f) Unit equipment must satisfy the applicable requirements of this subsection.

Individual emergency lighting units (ELUs) for emergency illumination must consist of:
- a rechargeable battery;
- a battery charging means;
- provisions for one or more lamps mounted on the equipment, or terminals for remote lamps, or both; and
- a relaying device arranged to energize the lamps automatically upon failure of the supply to the ELU.

Batteries used in unit equipment must be of suitable rating and capacity to supply and maintain at not less than $87^1/_2\%$ of the nominal battery voltage for the total lamp load associated with the unit for a period of at least $1^1/_2$ hours; or the unit equipment must supply and maintain not less than 60% of the initial emergency illumination for a period of at least 1½ hours. In addition, the batteries, whether acid or alkali type, must meet the requirements of emergency service.

ELUs must be permanently fixed in place (not portable) and have all wiring to each unit installed in accordance with the requirements of any of the wiring methods listed in NEC Chapter 3. As shown in **Fig. 2.2**, a flexible cord-and-plug connection is permitted, provided that the cord does not exceed 3 ft (914 mm) in length.

Legally required standby illumination heads that obtain power from an ELU, but are mounted remote from the unit, must be wired to the ELU by one of the wiring methods listed in NEC Chapter 3.

The branch circuit feeding the unit equipment must be the same branch circuit as that serving the normal lighting in the area and connected ahead of any local switches. An exception allows a separate circuit to feed the unit equipment if it is in a large uninterrupted area where at least three circuits from the same panel as the ELU feeds the lighting fixtures. The circuit to the unit equipment must be provided with a lock-on feature. This permission duplicates the one for emergency systems.

ARTICLE 701, PART D
LEGALLY REQUIRED STANDBY SYSTEMS — OVERCURRENT PROTECTION

Sec. 701-15 — Accessibility

The branch-circuit overcurrent devices in legally required standby circuits must be accessible to authorized persons only.

According to the definition in Article 100, this would require that it be guarded from approach by unauthorized persons by a locked door, or other effective means.

Sec. 701-17 — Ground-fault protection of equipment

The alternate source for legally required standby systems are not required to have ground-fault protection of equipment.

Not all auxiliary power sources are emergency or legally required items. Lost production or damage to goods are only considerations for installing an optional standby unit, provided life safety is not endangered by loss of prime power.

ARTICLE 702 — OPTIONAL STANDBY SYSTEMS

Having dealt with emergency systems in Chapter 1 and legally required standby systems in Chapter 2 (both of which are mandated by law), this chapter deals with the same type of equipment installed where safety is not the prime consideration. From an engineering standpoint such a system may be considered useful in preventing economic loss, for cogeneration purposes, or other reasons, but is not a legal requirement.

ARTICLE 702, PART A
OPTIONAL STANDBY SYSTEMS — GENERAL

Sec. 702-1 — Scope

Provisions of this article apply to the installation and operation of optional standby systems. The systems covered consist only of those that are permanently installed in their entirety, including the prime movers.

Sec. 702-2 — Optional standby systems

Systems that are intended to protect private or public business or property where life safety does not depend on the performance of the system are identified as being optional standby systems. They are intended to supply onsite generated power to selected loads either automatically or manually.

An FPN states that optional standby systems are typically installed to provide an alternate source of electric power for such facilities as industrial and commercial buildings, farms, and residences. In these facilities they typically serve loads such as:
- heating and refrigeration systems;
- data processing and communication systems; and
- industrial processes that when stopped during any power outage could cause discomfort, serious interruption of the process, damage to the product or process, or the like.

As this definition implies, optional standby systems are applied

Manufacturing facilities often contain computerized equipment with onboard backup supplies to carry them through a loss of power for a limited period of time. An optional standby system serves to preserve memory during prolonged power outages.

where failure of the normal electric power supply is not considered by the authority having jurisdiction to create a life-safety hazard. In such a situation, it is up to the owner of the facility to decide whether backup power is required for all or some of the equipment in order to prevent discomfort or economic loss.

Although not referred to in this article of the Code, NFPA 110 Sec. 2-3.5.4 classifies this type of standby system as Level 3. For further information on the subject, see Chapter 1 in this book.

Sec. 702-3 — Application of other articles

Except as modified by this article, all applicable articles of the NEC also apply to optional standby systems. As with emergency and standby systems, the provisions of this article are intended only to supplement or modify the requirements of Chapters 1 through 5 of the NEC. The rules in Article 702, thus, must be coordinated with those of the rest of the Code.

Sec. 702-4 — Equipment approval

All optional standby equipment must be approved for the intended use.

Although this equipment is applied at the discretion of the owner of the facility, rather than being mandated by the NEC, the equipment used for this purpose must be approved by the authority having jurisdiction. Misapplication of equipment can cause inadvertent hazards to the safety of personnel. Another reason for this requirement is that an early review could uncover details of the process that might not earlier have seemed to present a safety hazard requiring either emergency or legally required standby systems.

Sec. 702-5 — Capacity and rating

An optional standby system must have adequate capacity and rating for the supply of all equipment intended to be operated at one time. The equipment must be suitable for the maximum available fault current at its terminals.

Even though this equipment is installed only for the convenience of the owner, these Code rules must still be met because misapplication of equipment can cause equipment intended to serve as an optional standby system to become a safety hazard to personnel. Overloading and insufficient interrupting capacity are major causes of such catastrophic equipment failure.

This rule, however, then goes on to make clear that it is up to the owner/user of the optional standby system to select the loads to be connected to the system. If certain loads are required by law to be carried on the system, then it no longer qualifies as being an optional standby system.

Sec. 702-6 — Transfer equipment

Transfer equipment must be suitable for the intended use and so designed and installed as to prevent the inadvertent interconnection of normal and alternate sources of supply in any operation of the transfer equipment.

When located on the load side of branch-circuit protection, transfer equipment is permitted to contain supplementary overcurrent

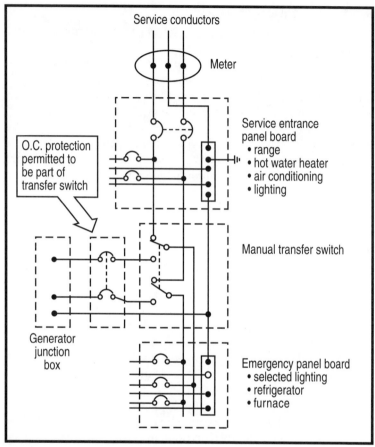

■ **Fig. 3.1.** An permanently mounted engine-generator set used in a residential occupancy to provide power during an electric utility power failure is classified as an optional standby power system. For such an application, this is a typical hookup. Equipment grounding is not shown.

protection having an interrupting rating sufficient for the available fault current that the generator can deliver.

A typical arrangement of a *permanently mounted* optional standby system installed in a single-family residence is shown in **Fig. 3.1**. A portable unit is not covered by this article.

Sec. 702-7 — Signals

Audible and visual signal devices must be provided (where practicable) to indicate the following:
- failure or malfunction of the optional standby source; and
- that the optional standby source is carrying a load.

This may seem like an unnecessary intrusion on the right of the owner of the equipment to decide how best to operate their system. But remember, this is often a borderline application where some types of failure of the equipment or the fact that an engine-generator or other source is operating may have an unanticipated effect on safety. Thus, the operational phrase here is, "where practicable." This is not a hard-and-fast requirement, but it should be implemented in order to provide the user with some valuable information. NFPA 110 does not set any requirements for signals for this type of equipment.

Sec. 702-8 — Signs

The following signs are to be posted.

- A sign must be placed at the service-entrance equipment indicating the type and location of onsite optional standby power sources.
- Where the grounded circuit conductor connected to the emergency source is connected to a grounding electrode conductor at a location remote from the emergency source, a sign must be posted at the grounding location that identifies all emergency and normal sources connected at that location.

These items echo those for emergency power and legally required standby power systems. In this case, unlike the requirement of Sec. 702-7 for signals, the signs are mandatory. Here a matter of safety is involved; persons working on the system must be aware of the presence of a second source of power. As with the other types of alternate power sources, however, unit equipment is exempted from the placard rule. It would be pointless to list every piece of unit equipment along with the major power sources. They pose virtually no safety threat, nor will they backfeed into the system.

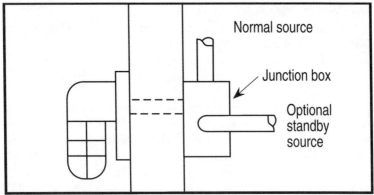

■ **Fig. 3.2.** In an installation containing an optional standby system, the normal and standby sources can be within a common enclosure. In addition, the two circuits can be run within a common raceway.

ARTICLE 702, PART B
OPTIONAL STANDBY SYSTEMS — CIRCUIT WIRING

Sec. 702-9 — Wiring optional standby systems

Wiring of an optional standby system is permitted to occupy the same raceways, cables, boxes, and cabinets with other general wiring.

Where it is desired (but is not required) to keep a piece of equipment (including lighting) in operation during a power outage, the NEC rule for separation of circuit wiring is relaxed. As shown in **Fig. 3.2,** both the normal and standby source wiring can both be within a common enclosure. This permission duplicates that allowed for legally required standby systems.

DIFFERENCE BETWEEN EMERGENCY AND STANDBY SYSTEMS

4.

One of the most misunderstood electrical term applied to circuitry is "emergency." Although all emergency systems (Chapter 1 of this book) are standby in character, there are many standby systems that are *not* emergency systems within the meaning of the NEC. These other systems, whether legally required (Chapter 2) or optional (Chapter 3), are subject to far different requirements. The NEC can be violated by applying rules that are not acceptable to use for emergency circuits.

Legal requirements

It is not permissible to pick and choose the rules of Article 700, 701, and 702 in order to fit a power source classified as a standby unit into the design of a system intended to supply the circuits of an emergency system. Once a standby power source is properly classified as an emergency system source, only the requirements in Article 700 can be utilized. Articles 701 and 702 must not be used to reduce the burden of stricter wiring techniques and greater cost.

A standby power source is either required by a statute or legally enforceable administrative regulation, or it is not. Emergency systems and legally required standby systems share a common thread in that both are legally required. If a standby system is *not* legally required, then it is *not* an emergency or legally required standby system, no matter how devastating the owner perceives a potential failure to be. In such cases, the owner's engineers must design adequate performance into the system; they cannot rely on the NEC to do that job for them.

Once it is established that the standby system is required by law, then it must be classified as *either* an emergency system or simply a required standby system. In some cases, the applicable regulation will stipulate the specific Code article, making the job easy. In some other cases, this determination can be difficult to make, since there is some overlap in the two articles. The most important criterion is

Factory automation complicates deciding whether equipment is required by the NEC to have an emergency system or legally required standby system as a backup source to the normal power, or whether the standby source is optional. In general, if dangerous conditions will be created by loss of power, then an emergency or legally required system is required. If only a loss of product is involved, then the owner may choose to include an optional standby system. A careful study of the process is required to make the correct determination.

---

the relative length of time that power can be interrupted without undue hazard.

The applications of true emergency systems are best expressed in FPN No. 3 following Sec. 700-1. It states:

> Emergency systems are generally installed in places of assembly where artificial illumination is required for safe exiting and for panic control in buildings subject to occupancy by large numbers of persons, such as hotels, theaters, sports arenas, health care facilities, and similar institutions. Emergency systems may also provide power for such functions as ventilation where essential to maintain life, fire detection and alarm systems, elevators, fire pumps, public safety communication systems, industrial processes where current interruption would produce serious life safety or health hazards, and similar functions.

There are three overlapping considerations, all of which relate to the basic question of how long an outage would be tolerable.

Nature of the occupancy is the first, with particular reference to the likely numbers of people that would be assembled at any one time. A large congregation of people in any single area is ripe for panic in the event of a fire, particularly if the area goes dark. Panic must be avoided, since it can develop with blinding speed and contribute to more deaths than the fire or other problem that caused it. Therefore, buildings with high-occupancy levels, such as high-rise buildings and large auditoriums, usually *require* emergency systems.

Criticality of the loads is the next. Generally, egress lighting and exit directional signs must be available at all times. In addition to lighting, signaling and communication systems, particularly those that are essential to public safety, must also be available with minimal interruption. Finally, some other loads, such as fire pumps or some ventilation systems essential to preserving life, will be unable to perform their intended function if they are disconnected for any length of time.

Danger to personnel during an outage is the third consideration. Some industrial processes, although not involving high-occupancy levels, are extremely dangerous to workers should power suddenly fail. An example would be an exothermic reaction in a vessel that could "run away" if sufficient cooling water is not supplied due to failure of power to the pumps.

To the extent reasonably possible, the applicable regulation will designate the areas (and/or loads) that must be served by the emergency system. One document widely used for this purpose is *NFPA 101, Code for Safety to Life from Fire in Buildings and Structures*. For example, Paragraph 28-2.9 requires all industrial occupancies to have emergency lighting for designated stairs, corridors, etc. that lead to an exit. There are exceptions for uninhabited operations and those that allow adequate daylight for all egress routes during production. A note in NFPA 101 advises authorities having jurisdiction (AHJs) to review large locker rooms, and laboratories using hazardous chemicals, to be sure that major egress aisles have adequate emergency illumination. Note that in such cases the AHJ is usually a building official and not an electrical inspector.

Provisions of Article 701 of the NEC apply to the installation, operation, and maintenance of legally required systems *other than those classed as emergency systems*. This class of standby system automatically supply power to selected loads in the event of failure of the normal source. The loads consist of circuits and equipment intended to supply, distribute, and control electricity to facilities for illumination and/or power when the normal electrical supply system is interrupted. These systems must be permanently installed in their entirety, including the power source.

The key to separating legally required standby from emergency systems is the length of time an outage can be tolerated. Sec. 701-2 (FPN) gives a good sense of typical loads served by such systems:

> Legally required standby systems are typically installed to serve loads, such as heating and refrigeration systems, communication systems, ventilation and smoke removal systems, sewerage disposal, lighting systems, and industrial processes, that, when stopped during any interruption of the normal electrical supply, could create hazards or hamper rescue or fire fighting operations.

These systems are less critical in terms of time for restoration, although they may yet be very critical in terms of long term environmental consequences, for example. They are also directed at the performance of selected electrical loads instead of the safe egress of personnel. For example, there are numerous rules requiring standby power for large sewage treatment facilities. In this case, the facility must remain in operation in order to prevent environmental problems, and some lighting may be required for this. Although critical, this is a different type of concern, allowing a longer time delay between loss and restoration than is permitted for lighting that is crucial to emergency egress.

Optional standby systems

Standby systems that are unrelated to life safety, regardless of how critical the owner views certain loads, are classified as Op-

Portable generators are not covered by the rules of Articles 700, 701, or 702.

tional Standby Systems. The provisions of Article 702 of the NEC apply to the installation and operation of these standby systems. Optional standby systems are intended to supply on-site generated power to selected loads either automatically or manually.

These systems, for the purpose of applying Article 702, consist only of those that are permanently installed in their entirety, including prime movers. A portable generator is *not* covered by this article; any load served by this source is covered by the general rules in the NEC.

There is a big difference between true emergency circuits covered by Article 700 and those that are served by standby or optional sources per Article 701 and 702. Care must be exercised to select the correct article that apply to a system in question.

Equipment approval. All the equipment used for emergency systems per Sec. 700-3 must be approved for such use. Where listed equipment is available, most inspectors will insist on emergency equipment that is listed for such applications by a qualified testing laboratory.

This is different from legally required standby and optional standby systems. Sec. 701-4 and Sec. 702-4 merely require that this equipment be *approved* for its intended use.

Test and maintenance procedures. On emergency systems, the AHJ must conduct or witness a test on the complete system upon installation. This test must include the maximum anticipated load during operation of the emergency system. Such tests must also be performed and witnessed periodically following the initial test to completely satisfy Sec. 700-4(a).

Sec. 701-5(a) on legally required standby systems requires the AHJ to conduct or witness such tests upon installation, but does *not* mention periodic testing afterwards. Art. 702 for optional systems does not even require the acceptance test.

Capacity and rating. An emergency system per Sec. 700-5(a) must be sized with enough capacity to adequately start and carry all loads *simultaneously*. This is different from legally required standby systems, which per Sec. 701-6 must be sized with enough capacity to start and carry the loads of all equipment *intended to be operated at one time*. Optional standby systems (per Sec. 702-5) only have to be sized to carry loads that are selected by the user.

In general, the equipment used in these systems must be suitable for maximum available fault current available at its line terminals.

Notice that careful consideration must be given to the requirements involved where the power source is used for peak load shaving. Emergency generating capacity is allowed to be used for this purpose, but *only* where automatic means are provided to assure adequate power to the emergency loads. After their function is assured, then the legally required standby loads can be energized. Finally, optional standby loads can be served when the other systems have been restored.

Legally required standby systems can also supply optional standby loads as long as automatic selective load pickup and load shedding is employed to assure that, when needed, legally required loads are given first priority.

No restrictions are placed on the use of optional standby systems for peak load shaving or other purposes.

Transfer equipment. Here the rules are similar for both Articles 700 and 701. The transfer equipment must prevent the inadvertent interconnection of normal and alternate supplies, to avoid hazards

Transfer switches are essential parts of an electrical system including alternate sources of power.

to personnel working on the line side of the equipment. A reference, however, is made to Sec. 230-83, which includes an exception permitting the use of approved "closed-transition switches."

For optional standby systems only, Sec. 702-6 provides that transfer equipment located on the load side of branch-circuit protection is permitted to incorporate supplementary overcurrent protection with an interrupting rating sufficient for the available generator fault current.

A typical 7500W generator has an available fault current in the 360A range. Conventional branch-circuit devices suitable for 10,000A and over required on utility sources need *not* be applied in these cases. Although these small transfer switches are often used with portable generators that are not within the scope of Art. 702, at times they are used with permanently installed systems.

Signals. In each case, the following requirements are made conditional by the words, "where practicable." For emergency systems, Sec. 700-7(b) requires a signal, both audible and visual, to indicate when the battery is carrying load. For legally required standby systems, Sec. 701-8 requires a signal to indicate that the standby source is carrying the load. Since optional standby systems supply loads selected by the owner of such systems, the only signals required by Sec. 702-7 are to indicate derangement of the system and to verify the system power source is carrying the load.

Wiring. The general rule of Sec. 700-9 requires that the wiring for emergency systems be *kept entirely independent* of the regular wiring used for lighting and that it thus needs to be in separate raceways, cables, and boxes. The concept of this rule is to ensure that where faults occur on the regular wiring, they will not affect the emergency system wiring, for such wiring will be in separate enclosures.

There are exceptions to Sec. 700-9. They all apply to obvious terminations at dual source equipment, and *never* permit emergency circuits to travel in common raceways with other power circuits. Note that it is not allowed for taps to be made within a transfer switch to feed non-emergency loads.

This section also requires that emergency circuits be so located as to minimize the possibility of failure due to flooding, fire, vandalism, and other adverse conditions. This is stronger language than "consideration should be given" to these problems. Some designers, when confronted by the inspector, have sometimes asserted that they indeed had considered the problem and thus complied with the rule.

This area is probably the greatest area of difference from the other standby system rules. Sec. 701-10 and 702-9 *do* permit the wiring of legally required and optional standby systems to occupy the same raceways, cables, boxes, and cabinets with the normal wiring circuits. In addition, the requirement to avoid flooding hazards, etc., is *not* repeated in the other articles.

Transfer time. Sec. 700-12 specifies that the normal-to-emergency transfer *must not exceed 10 sec.* This is another important area where the rules for emergency systems are more severe than for the other systems. Sec. 701-11 requires legally required standby power to be available within the time required for the application, but sets a limit of *not more than 60 sec* to be completely on the line. Optional standby systems do *not* have a time limitation to come on line to provide auxiliary power for selected loads. Sec. 702-2 recognizes a manual transfer, which could obviously take place at any time.

Generator sets that require more than 10 sec. to develop power are acceptable for use in emergency systems providing an auxiliary power supply will energize the emergency system until the generator

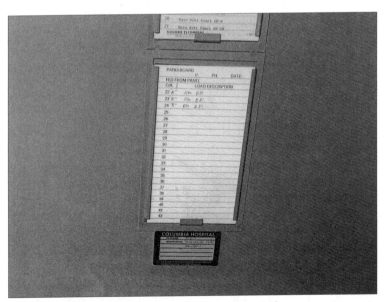

Panelboard directories must be completed with care. When unit equipment is connected to a circuit, the circuit *must* be identified in the listing.

can pick up the load [Sec. 700-12(b)(5)]. Sec. 701-11 does *not* list such a requirement for legally required standby systems, and similarly there is *no* comparable rule for optional standby systems in Article 702.

Unit equipment. A requirement in Sec. 700-12(f) for emergency systems mandates that the branch circuit feeding the unit equipment be *specifically identified* as such at the distribution panel. A typical branch circuit for lighting in a hallway will usually be identified as "hallway lighting." Without experimenting with the various hallway lighting circuits, the inspector has no way of knowing which circuit carries the unit equipment. In violation of this section, unit equipment has often been connected to circuits other than the local lighting circuit. This identification requirement makes the source obvious and assists proper enforcement. This rule is *not*, however, found in Sec. 701-11(f) for legally required systems, and does *not* apply in Art. 702 either.

Ground-fault protection. Per Sec. 700-26, an emergency generator disconnect is not required to have ground-fault protection of

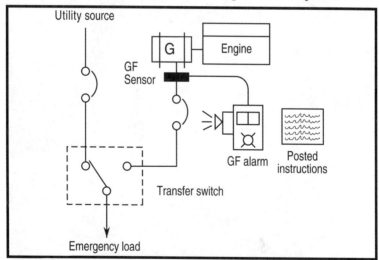

■ **Fig. 4.1.** Presence of a ground fault on an emergency system must be sensed and alarmed. In addition, instructions must be posted at the alarm location detailing the response required.

equipment (GFPE) against arcing faults. This provision means that Sec. 230-95 and Sec. 215-10, which normally require GFPE protection for 1000A and larger service disconnects (and feeder disconnects where not protected by service GFPE), do *not* apply to an emergency system disconnect. This increases the reliability of the system.

A sensor to detect a ground fault, however, must be installed and connected so as to provide a *signal* to indicate such a condition may exist [Sec. 700-7(d)]. Instructions as to how personnel should respond to such a fault must be provided at the sensor location. This is shown in **Fig. 4.1**.

Although Sec. 701-17 does not require ground-fault protection for legally required standby systems as well, there is *no* comparable rule for sensing and indicating a ground-fault. Article 702 does *not* address the issue. Since the other rules of the Code apply, per Sec. 702-3, the *normal* requirements for GFPE apply to these optional systems.

Lighting circuit reliability. Another key difference between the articles centers on specific requirements assuring lighting reliability that *are* included in Article 700, but *not* addressed in Articles 701 and 702 at all. Sec. 700-15, for example, requires that no appliances or other loads be connected to an emergency lighting circuit. Those circuits are completely reserved for their emergency functions.

To further increase reliability, Sec. 700-16 includes requirements for redundancy and continuity of required emergency lighting. These provisions are also not repeated in the other articles. This section also defines what is meant by emergency illumination, expressly including illuminated exit signs, egress lighting, and other lighting as required. Furthermore, no single lighting element failing can put any space required to be illuminated in total darkness.

Another requirement for emergency systems addresses the hot restrike problem with conventional HID lighting sources. In these cases, the emergency system must stay on *until normal lighting is restored*, as opposed to merely the restoration of normal power.

Article 700 also has requirements for lighting controls that are not mandated for legally required standby or optional standby systems.

Example

One example of gray areas between the various systems is a high-rise building with a 480Y/277V emergency lighting system powered by a generator located in its penthouse. The penthouse switchroom also has a small transformer and a 208Y/120V panel for selected 120V emergency loads. Years later, the first floor of the building is retrofitted as a security office, which involves connecting the central annunciator system (with partial battery back-up) for fire alarm systems in several buildings. The only feasible route from the penthouse to the first floor involves using existing raceways that enclose other power circuits.

There is no question that the annunciator and other fire-alarm system circuitry on the load side of the system power supply qualify as an emergency system, and can not be run in other raceways. The question concerns the status of the branch circuit feeding that power supply. Sec. 1-5.2.3 of *NFPA 72, National Fire Alarm Code* allows both emergency and legally required standby systems to provide the power for these systems, provided the continuity requirements of Sec. 1-5.2.6 are met. Battery back-up provides the required continuity, therefore, the branch circuit qualifies as being legally required and not emergency. This, in turn, means that an existing conduit riser to the penthouse can be used for this circuit.

Remember, it is the *essential elements* that count in deciding whether a system is emergency or not. In this example, the applicable building code identifies the protected *occupancies* as being *critical* and requiring the protection. Therefore, a *legal requirement* is in force on those buildings. The *load* was similarly identified as *critical*, due to the *personnel hazard* of leaving fire-alarm systems nonfunctional. On the supply side of the power supply, however, a *longer outage could easily be tolerated* due to the battery back-up. Although legally required, this standby circuit is not an emergency circuit.

ARTICLE 705 — INTERCONNECTED ELECTRIC POWER PRODUCTION SOURCES

5.

Increasingly, power generation units are being installed that can serve as both a backup to the normal source of electricity, to provide a method of providing power for peak shaving, and similar strategies for reducing utility bills. Some of these units can generate more power than needed by the facility and, thus, are available to feed power into the utility grid. In other instances, surplus steam from boilers can be used to drive a turbine-generator. Such a cogeneration scheme also includes the possibility of feeding power back into the utility grid.

To achieve this type of operation, it is necessary to run the onsite power-generation system in parallel with the utility or other source of primary power. This article addresses the requirements for such operations.

Sec. 705-1 — Scope

This article covers installation of one or more electric power production sources operating in parallel with a primary source or

Turbine generators driven by surplus steam produced in process industries, such as paper mills, can feed power back into an electric utility's grid, thereby paying for fuel costs of generating the steam.

sources of electricity. An FPN explains that the primary source can be a utility supply, an on-site electric power source, or other source.

Sec. 705-2 — Definition

For purposes of this article, the following definition applies.

An *interactive system* is an electric power production system that is operating in parallel with, and capable of delivering energy to, an electric primary source supply system.

This definition makes clear that this article applies to cogeneration systems where onsite power sources can generate in parallel with the electric utility. It also applies, however, to engine-generators and other onsite sources that are tied into a main facility power system derived from other onsite generating equipment.

Sec. 705-3 — Other articles

Besides the articles dealing with emergency, legally required standby, and optional standby systems, the interconnected electric power production sources must comply with NEC Article 445 (Generators) and Article 690 (Solar Photovoltaic Systems) when applicable.

Multiple sets of generating equipment can be identified as a group rather than listing each separately.

Sec. 705-10 — Directory

A permanent plaque or directory, identifying all electrical power sources on or in the premises, must be installed at each service equipment location and at locations of all electric power production sources capable of being interconnected.

An exception to the rule is made for situations where there are a large number of power production sources. The directory can identify the sources by groups. For instance, if six engine-generator sets operating in parallel with the utility system are installed in a machine room, it would only be necessary to post a directory specifying the location of the group at the utility service and machine room rather than listing each individual unit.

Sec. 705-12 — Point of connection

Outputs of electric power production sources must be interconnected at the service disconnecting means. A reference is made to Sec. 230-82, Ex. 6. This reference is to the rule that equipment must not be connected to the supply side of the service disconnect means. Exception 6, however, *allows* interconnected electric power production sources (and photovoltaic systems) to be connected on the supply side in order to prevent both power sources from being interrupted when the main service disconnect is opened.

Note the use of the word "at" in Sec. 705-12. It does not specify whether the electric power production must be ahead or downstream of the main disconnect. As shown in **Fig. 5.1** on the next page, depending upon the system design either is acceptable. Be aware, however, that **Sec. 705-40** mandates that the electric power production soures must be automatically disconnected upon the loss of the primary source. More about this later.

This flexibility in determining the connection point is confirmed by the exceptions to the requirement of Sec. 705-12.

<u>Exception No. 1</u>. Outputs of electric power production sources are permitted to be interconnected at a point or points elsewhere on the premises if the system qualifies as an integrated electric system, and it incorporates protective equipment in accordance with all applicable sections of Article 685.

59

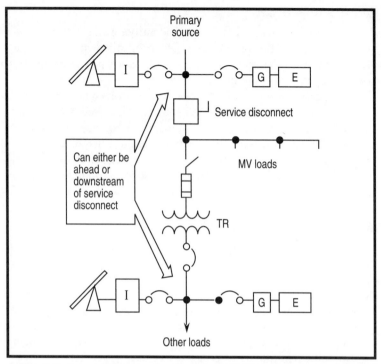

■ **Fig. 5.1.** An exception is given in Sec. 230-82 to permit connection of interconnected electric power production sources on the line side of the service disconnecting means.

Sec. 685-1 identifies an integrated electrical system as being a unitized segment of an industrial wiring system where all of the following conditions are met:
• an orderly shutdown is required to minimize personnel hazard and equipment damage;
• the conditions of maintenance and supervision assure that qualified persons will service the system; and
• effective safeguards, acceptable to the authority having jurisdiction, are established and maintained.

Article 685 only applies to industrial locations and the article does not define protective equipment that will allow the systems to operate satisfactorily in parallel with another source of power. The nearest to such a listing of requirements is **Sec. 685-2**, but this only refers to needs for ground-fault protection, overcurrent protection, grounding, disconnecting means, etc.

Exception No. 2. Outputs are permitted to be interconnected at a point or points elsewhere in the premises where all the following conditions are met:

• the aggregate of non-utility sources of electricity has a capacity in excess of 100kW, or the service is above 1000V;

• the conditions of maintenance and supervision assure that qualified persons will service and operate the system; and

• safeguards and protective equipment are established and maintained.

Sec. 705-14 — Output characteristics

The output of a generator or other electric power production source operating in parallel with an electric supply system must be compatible in voltage, wave shape, and frequency to that of the

Lineups of metalclad gear provide a centalized point at which all sources of electric power can be interconnected, monitored, and controlled.

system to which it is connected.

An FPN explains that "compatible" as used in this rule does not necessarily mean that the primary source wave shape must be matched. The output of a UPS system is a synthesized AC wave, which will not exactly match that of an onsite generator or utility line.

Sec. 705-16 — Interrupting and withstand rating

Consideration must be given to the contribution of fault currents from all interconnected power sources for the interrupting and withstand ratings of equipment on interactive systems.

This differs from other articles requiring that the equipment must be "suitable" rather than just consideration be given to this point. The intent, however, clearly is that such is the case here also.

Sec. 705-20 — Disconnecting means, sources

Means must be provided to disconnect all ungrounded conductors of an electric power production source(s) from all other conductors.

A reference is made to Article 230, which covers services. Part F of that article deals with the required disconnecting means. Most of the applicable rules listed there are also included in Sec. 705-22.

Sec. 705-21 — Disconnecting means, equipment

Means must be provided to disconnect equipment, such as inverters or transformers, associated with a power production source, from all ungrounded conductors of all sources of supply.

An exception to this rule is made for equipment intended to be operated and maintained as an integral part of a power production source exceeding 1000V.

When interconnected power sources are present, safety requires that equipment connected to the system have a means of totally isolating them from energy sources to allow maintenance to be carried out without danger to personnel.

Sec. 705-22 — Disconnect device

Disconnecting means for ungrounded conductors must consist of a manually or power operable switch(es) or circuit breaker(s) that are:
- located where accessible;
- externally operable without exposing the operator to contact with live parts and, if power operable, of a type that can be easily opened by hand in the event of a power supply failure;
- plainly indicating whether it is in the open or closed position; and
- having ratings not less than the load to be carried and the fault current to be interrupted.

For disconnect equipment energized from both sides, a marking must be provided to indicate that all contacts of the disconnect equipment may be energized.

An FPN explains that in parallel generation systems, some equipment, including knife blade switches and fuses, are likely to be energized from both directions. A reference is made to **Sec. 240-40**. This section requires that disconnect switches be provided for fuses.

A second FPN states that interconnection to off-premises primary source could require a visibly verifiable disconnecting device.

Sec. 705-30 — Overcurrent protection

Conductors must be protected in accordance with **Article 240**. Equipment overcurrent protection must also be in accordance with that article.

In general, equipment and conductors connected to more than one electrical source must have a sufficient number of overcurrent devices located so they will provide protection from all sources. This requirement is similar to that for disconnecting means. Equipment must be able to be isolated from all sources of supply to insure that it can be worked on safely, and also protected by overcurrent devices against backfeed from downstream sources.

Sections of other articles must also be satisfied, including the following.

• Generators must be protected in accordance with **Sec. 445-4**. This will be discussed later in this chapter.

• Solar photovoltaic systems must be protected in accordance with **Article 690**. The principal requirements there are similar to those listed in Sec. 705-30.

• Overcurrent protection for a transformer with a source(s) on each side must be provided in accordance with **Sec. 450-3** by considering first one side of the transformer as the primary, then the other side of the transformer as the primary.

As well as applying to an individual single-phase or 3-phase transformer, the rules of Secs. 450-3 also apply to a bank of single-phase transformers connected to operate as a single 3-phase unit. Besides the following rules described here, there are many other rules in this section that serve to modify the sizing requirements for overcurrent protection of transformers. For a more full discussion of these items see *EC&M's* book entitled, *Understanding NE Code Rules on Transformers*.

Sec. 450-3(a) says that transformers must have primary and secondary protective devices rated or set at no more than the values of transformer rated currents noted in NEC **Table 450-3(a)(1)**. The values given in that table are reproduced in **Fig. 5.2**. Electronic fuses that can be set to open at a specific current are to be treated similar

	Maximum Rating or Setting for Overcurrent Device				
	Primary		Secondary		
	Over 600 Volts		Over 600 Volts		600 Volts or Below
Transformer Rated Impedance	Circuit Breaker Setting	Fuse Rating	Circuit Breaker Setting	Fuse Rating	Circuit Breaker Setting or Fuse Rating
Not more than 6%	600%	300%	300%	250%	125%
More than 6% and not more than 10%	400%	300%	250%	225%	125%

■ **Fig. 5.2.** Maximum rating or setting for overcurrent protective devices for transformers connected to a system over 600V. Derived from NEC Table 450-3(a)(1).

to circuit breakers. Note that this table applies only to transformers having a primary operating at over 600V.

EXAMPLE: Assume in **Fig. 5**.3 that the primary current is approximately 44A, and the secondary current is approximately 482A. Size the overcurrent protection of primary and secondary.

ANSWER: Based on these currents, the maximum rating of the primary fuses for a system operating at over 600V is:

$$44 \times 6 = 264A$$

and the maximum setting of the circuit breaker protecting the secondary with an operating voltage over 600A is:

$$482 \times 3 = 1446A.$$

An exception to the general rule says that if the required fuse rating or circuit breaker setting does not correspond to a standard rating or setting, the next higher standard rating or setting is

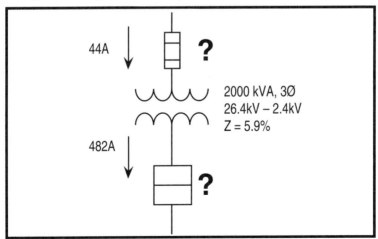

Fig. 5.3. Sizing primary and secondary overcurrent protective devices for transformers operating at over 600V.

permitted.

In this example, 300A fuses can be used for the primary. This is based on the standard ampere ratings listed in NEC **Sec. 240-6(a)**. If the secondary circuit breaker has only a fixed trip, then a 1600A rating would seem to be the maximum allowed per the listing in **Sec. 240-6(a)**.

The example assumed a medium voltage secondary operating voltage. The calculation of the maximum rating of the circuit breaker if the secondary voltage was 600V or less are somewhat different. Assuming the impedance is not more than 10%, the secondary amperes would be permitted to be multiplied by a factor of only 1.25. As an example, for a 3-phase transformer with a 480V nominal secondary voltage and an impedance of 5.75% having a 600A rated secondary current, the calculated maximum size overcurrent device is:

$$600 \times 1.25 = 750A.$$

Another case to be considered is when the interconnected electric power production source is located on the downstream side of the secondary overcurrent protective device of Fig. 5.3. Here the secondary breaker can be energized from both directions. The rule is that first one side of the transformer and then the other is to be considered as the primary. The problem is that if the secondary side of the transformer is 600V or below, Table 450-3(a)(1) does not apply because there is no column for primaries at that voltage. In turn, the rules for sizing overcurrent devices for transformers operating at 600V or less also do not apply to a step-up transformer. The only solution seems to be to stay with the 125% rating of the secondary (now primary) device.

In an actual system, however, such a situation will rarely be presented. Sec. 705-40 requires that upon loss of the primary source, the electric power production source is to be disconnected for the ungrounded conductors of the primary source. In Fig. 5.1, this could be accomplished by opening the breaker on the transformer secondary or from the generator.

If the generator or other source must provide power to the internal loads during this power outage, then only the transformer secondary breaker could be opened. However, the generator is no longer a power production source covered by Article 705, but rather an auxiliary source covered by Articles 700, 701, or 702 during this period of operation.

Sec. 450-3(b) contains a set of rules applying to transformers connected to a system operating at a voltage of 600V or less. The general rule on sizing the overcurrent device on the primary of these transformers is that it must be set or rated at not more than 125% of the rated primary current of the transformer.

There are several exceptions to this rule dependent upon the rated primary current of the transformer. The one that would apply in this situation is: If it is 9A or more, and the 125% does not correspond to a standard rating of a fuse of nonadjustable circuit breaker, the next higher standard rating given in Sec. 240-6 is permitted.

A transformer whose primary is connected to a system operating at 600V, nominal, or less, is not required to have an overcurrent protective device on its primary if:

- it has an overcurrent device on its secondary side rated or set at not more than 125% of the rated secondary current of the transformer; and
- the overcurrent device protecting the feeder to the transformer primary is rated or set at not more than 250% of the rated primary current of the transformer.

Sec. 705-32 — Ground-fault protection

Rules that require GFP are to be found in Sec. 230-95. There it says that solidly grounded wye electrical services having a disconnect rated at 1000A or more and a voltage of more than 150V to ground (but not exceeding 600V phase-to-phase), are required to have ground-fault protection. The GFP must cause the service disconnecting means to open all ungrounded conductors of the faulted circuit.

Where ground-fault protection is used, the output of an interactive system must be connected to the supply side of the ground-fault

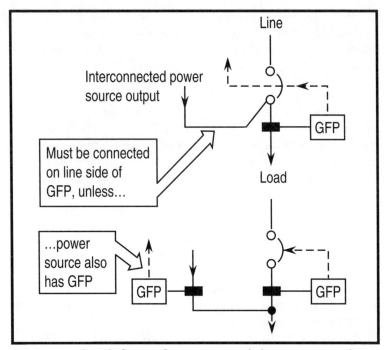

■ **Fig. 5.4.** The rule for GFP for interconnected electric power production sources.

protection (GFP).

An exception says that the connection is permitted to be made to the load side of the GFP provided there is also GFP for equipment from all ground-fault current sources. This concept is shown in **Fig. 5.4**.

There are several FPNs to Sec. 230-95 that offer help in interpreting the requirements for ground-fault protection. One of them recognizes that GFP must be desirable for service disconnecting means rated less than 1000A on solidly grounded wye systems. Another says GFP that opens the service disconnecting means will afford no protection from faults on the *line side* of the protective element; it only limits damage to conductors and equipment on the *load side* in the event of an arcing ground fault on the load side of the protective element.

Most importantly for cases involving interconnected electric power production sources, one of the FPNs states that where GFP is provided for the service disconnecting means and interconnection is made with another supply system by a transfer device, means or devices could be needed to assure proper ground-fault sensing of the ground-fault protective equipment. This would require relaying to prevent transfer to the auxiliary source of power if a ground fault is sensed.

The ground-fault protection system must be performance tested at the site when first installed. The test must be conducted in accordance with instructions that are provided with the equipment. A written record of this test must be made and is required to be available to the authority having jurisdiction.

Sec. 705-40 — Loss of primary source

Upon loss of the primary source, an interconnected electric power production source must be automatically disconnected from all ungrounded conductors of the primary source and must not be reconnected until the primary source is restored.

An FPN notes that risks to personnel and equipment associated with the primary source could occur if an interactive electric power production source can operate independently of the primary source. Special detection methods could be required to determine that a primary source supply system outage has occurred, and the extent of the electric power production source that must be automatically disconnected. This largely depends upon the design of the overall power distribution system.

When the primary source supply system is restored, special detection methods might be required to prevent power production sources from reconnecting while they are out-of-phase. Rotating equipment tied to the system can be severely damaged by the high torques resulting from the high voltages produced by out-of-phase reclosing.

Most of the requirements stated in the FPN are provided in automatic transfer switches. Reconnection out of phase is usually prevented by allowing sufficient time delay before retransfer once the primary source voltage has been reestablished.

A second FPN is a safety warning that says induction generating equipment on systems with significant capacitance can become self-excited upon loss of the primary source and experience severe overvoltage as a result.

This situation can be prevented by proper overvoltage protective relaying of the induction generating equipment.

Sec. 705-42 — Unbalanced interconnections

A 3-phase electric power production source must be automatically disconnected from all ungrounded conductors of the interconnected systems when one of the phases of that source opens.

An exception to this rule is made for an electric power production source providing power for an emergency or legally required standby system. In these cases, it is preferred to continue to carry the loads connected to the remaining phases rather than trying to minimize potential damage to equipment caused by single phasing.

Sec. 705-43 — Synchronous generators

A parallel system of synchronous generators must be provided with the necessary equipment to establish and maintain a synchronous condition.

Unlike induction generators, synchronous generators are not held in synchronism with a common source. Thus, where two or more are to be paralleled, some means must be provided to synchronize them as they are brought onto the line, and to maintain them synchronized as they run.

Induction generators are simple to parallel because they all synchronize with the source of field excitation (usually the utility service).

Sec. 705-50 — Grounding

Interconnected electric power production sources must be grounded in accordance with Article 250.

By the definition given in **Sec. 250-5(d), FPN No. 1,** an interconnected electric power production source is not considered to be a separately derived source if the neutral of the onsite generator is solidly interconnected to a service-supplied system neutral. Thus, the requirements of **Sec. 250-26** do not apply. The reference, thus, is mainly to Parts D, E, F, and G that deal with equipment grounding and bonding. For a more detailed explanation of grounding requirements, see *EC&M's* book entitled, *Understanding NE Code Rules on Grounding and Bonding.*

An exception is made for DC systems connected through an inverter directly to a grounded service. In that case, other methods that accomplish equivalent system protection and that utilize equipment listed and identified for the use is permitted.

The wording of this exception ties in closely with the exception to **Sec. 250-22** about the point where the grounding connection is to be made on a premises-located DC system.

Auxiliary power generators must, in addition to the rules of Articles 700, 701, 702, or 705, also meet the requirements of Articles 445 and 250.

ARTICLE 445 (GENERATORS) AND ARTICLE 250 (GROUNDING)

6.

NEC rules that specifically apply to generators, which are sources of power permitted in Articles 695, 700, 701, 702, and 705, are contained in Article 445. The requirements of this article, thus, must also be satisfied when installing an auxiliary power system.

Article 250 contains certain grounding rules that apply particularly to portable and vehicle-mounted generators. These are included in this chapter. Other rules of grounding included in that NEC article apply to all electrical systems and equipment. These are critical to the safe operation of any auxiliary power system and also must be adhered to. Listing all the applicable requirements here is beyond the scope of this book. For more information on this phase of the requirements, see *EC&M's* book entitled, *Understanding NE Code Rules on Grounding*.

Engine-driven generators are frequently used to provide emergency or standby power, or as part of an interconnected power production source. They must meet the requirements of Article 445.

Sec. 445-1 — General

Generators and their associated wiring and equipment must comply with applicable provisions of Articles 230, 250, 695, 700, 701, 702, and 705.

Sec. 445-2 — Location

Generators must by of a type suitable for the locations in which they are installed. They must also meet the requirements for motors in **Sec. 430-14**. The cited section refers to the location of motors. It states:

> (a) **Ventilation and Maintenance.** Motors shall be located so that adequate ventilation is provided and so that maintenance, such as lubrication of bearings and replacing of brushes, can be readily accomplished.
>
> (b) **Open Motors.** Open motors having commutators or collector rings shall be located or protected so that sparks cannot reach adjacent combustible material, but this shall not prohibit the installation of these motors on wooden floors or supports.

Generators installed in hazardous (classified) locations as described in Articles 500 through 503, or in other locations as described in Articles 510 through 517 must meet the requirements listed there. For more information on these see *EC&M's* book entitled, *Understanding NE Code Rules on Hazardous Locations*.

Articles 520, 530, and 665 are also referred to. Generators located in the facilities covered by these articles must also comply with the applicable provisions given there.

Article 520 deals with theaters, audience areas of studios, etc. **Sec. 520-8** simply refers to compliance with Article 700 for emergency systems.

Article 530 deals with movie and TV studios, etc. **Sec. 530-52** prohibits the location of portable electric equipment (which could include generators) in cellulose nitrate film storage vaults. **Sec. 530-63** requires that 3-wire DC generators have protection consisting of overcurrent devices with an amp rating or setting in accordance with the generator amp rating. Single- or double-pole overcurrent devices are permitted, and no pole or overcurrent coil is required in

Conversion equipment needed to provide power to induction or dielectric heating systems are covered by the rules of Article 665.

the neutral lead (whether grounded or ungrounded).

Article 665 deals with induction and dielectric heating equipment. Part C of the article covers motor-generator equipment. This is described in **Sec. 665-40** as including:

> ... all rotating equipment designed to operate from an ac or dc motor or by mechanical drive from a prime mover, producing and alternating current of any frequency for induction and/or dielectric heating.

Part D of the article covers equipment other than motor-generators. This equipment is described in **Sec. 665-60** as consisting of:

... all static multipliers and oscillator-type units utilizing vacuum tubes and/or solid-state devices. The equipment shall be capable of converting ac or dc to an ac frequency suitable for induction and/or dielectric heating.

The requirements for application of M-G sets and other equipment to such specialized applications as induction and dielectric heating equipment are very extensive and beyond the scope of this book. Refer to the appropriate Code parts that apply.

Article 525 (carnivals, circuses, fairs, and similar events) is not referred to, but contains requirements for generators used in these applications. Sec. 525-10(b)(2) requires that generators comply with the requirements of Article 445. Sec. 525-22 says that all equipment requiring grounding must be grounded by an equipment grounding conductor, which must be bonded to the system grounded conductor at the generator or first disconnecting means supplied by the generator. This connection is prohibited from being made on the load side of disconnecting means.

Sec. 445-3 — Marking

Each generator must be provided with a nameplate giving the makers's name, the rated frequency, power factor, number of phases (if AC), the rating in kW or kVA, the normal volts and amperes corresponding to the rating, rated rpm, insulation system class and rated ambient temperature or temperature rise, and time rating.

Sec. 445-4 — Overcurrent protection

This section presents the rules for protecting generators from overcurrents. There is, however, an exception to the following requirements for overcurrent protection that applies to all the types of generators. It says that where deemed by the authority having jurisdiction to be a generator that is vital to the operation of an electrical system and should operate to failure to prevent a greater hazard to persons, the overload sensing device(s) are permitted to be connected to an annunciator or alarm supervised by authorized personnel instead of interrupting the generator circuit.

(a) Constant-voltage generators, except AC generator exciters,

must be protected from overloads by inherent design, circuit breakers, fuses, or other acceptable overcurrent protective means suitable for the conditions of use.

(b) Two-wire generators are permitted to have overcurrent protection in one conductor only if the overcurrent device is actuated by the entire current generated other than the current in the shunt field. The overcurrent device must not open the shunt field.

(c) 65V or less. Generators operating at 65V or less and driven by individual motors are considered as protected by the overcurrent device protecting the motor if these devices will operate when the generators are delivering not more than 150% of their full-load rated current.

(d) Balancer sets. Two-wire DC generators used in conjunction with balancer sets to obtain neutrals for 3-wire systems (**Fig. 6.1**) must be equipped with overcurrent devices that will disconnect the 3-wire system in case of excessive unbalancing of voltages or currents.

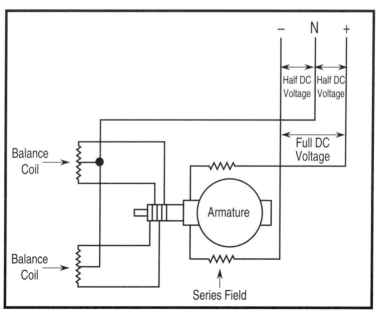

■ **Fig. 6.1.** A typical balancer set covered by Article 445.

(e) Three-wire DC generators, whether compound or shunt wound, must be equipped with overcurrent devices, one in each armature lead, and so connected as to be actuated by the entire current from the armature. Such overcurrent devices must consist either of a double-pole, double-coil circuit breaker, or of a 4-pole circuit breaker connected in the main and equalizer leads and tripped by two overcurrent devices, one in each armature lead. Such protective devices must be so interlocked that no one pole can be opened without simultaneously disconnecting both leads of the armature from the system.

Sec. 445-5 — Ampacity of conductors

The ampacity of the phase conductors from the generator terminals to the first overcurrent device must not be less than 115% of the nameplate current rating of the generator. Two exceptions to this section apply principally to this requirement.

Exception No. 1. Where the design and operation of the generator prevent overloading. In this case, the ampacity of the conductors are permitted to be not less than 100% of the nameplate current rating of the generator.

Exception No. 2. Where the generator manufacturer's leads are connected directly to an overcurrent device that is an integral part of the generator-set assembly.

It is also permitted by Sec. 445-5 to size the neutral conductors in accordance with **Sec. 220-22**. The rule there assumes that the feeder neutral load is the maximum unbalance of the load. In turn, the maximum unbalanced load is the maximum net computed load between the neutral and any one ungrounded conductor, except that the load thus obtained is to be multiplied by 140% for 3-wire 2-phase, or 5-wire 2-phase systems.

For a feeder supplying household electric ranges, wall-mounted ovens, counter-mounted cooking units, and electric dryers, the maximum unbalanced load can be considered to be 70% of the load on the ungrounded conductors.

For 3-wire DC, single-phase AC, 3-phase 4-wire, 2-phase 3-wire, or 2-phase 5-wire systems, a additional demand factor of 70%

is permitted for that portion of the unbalanced load in excess of 200A.

There must be no reduction of the neutral capacity for that portion of the load that consists of nonlinear loads supplied from a 3-phase 4-wire wye- connected system, nor the grounded conductor of a 3-wire circuit consisting of two-phase wires and the neutral of a 4-wire, 3-phase, wye-connected system. An FPN explains that a 3-phase 4-wire power system used to supply power to nonlinear loads might necessitate that the power system design allow for the possibility of high harmonic neutral currents.

Similarly, there is to be no reduction in ampacity of the grounded conductor of a 3-wire circuit consisting of two phase wires and the neutral of a 3-phase 4-wire wye-connected system.

Sec. 250-23(b) is also cited in Sec. 445-5 as the source for determining the sizing of generator conductors that must carry ground-fault currents.

The requirement there is that where an AC system operating at less than 1000V is grounded at any point, the grounded conductor must not be smaller than the required grounding electrode conductor specified in **Table 250-94**. The values there have been reproduced in **Fig. 6.2**. For phase conductors larger than 1100 kcmil

Largest Service-Entrance Conductor or Equivalent Area for Parallel Conductors		Size of Grounding Electrode Conductor	
Copper	Aluminum or Copper-Clad Aluminum	Copper	Aluminum or Copper-Clad Aluminum
2 or smaller	1/0 or smaller	8	6
1 or 1/0	2/0 or 3/0	6	4
2/0 or 3/0	4/0 or 250 kcmil	4	2
Over 3/0 thru 350 kcmil	Over 250 kcmil thru 500 kcmil	2	1/0
Over 350 kcmil thru 600 kcmil	Over 500 kcmil thru 900 kcmil	1/0	3/0
Over 600 kcmil thru 1100 kcmil	Over 900 kcmil thru 1750 kcmil	2/0	4/0
Over 1100 kcmil	Over 1750 kcmil	3/0	250 kcmil

■ **Fig. 6.2.** Size of grounding electrode conductor for AC systems. Derived from NEC Table 250-94.

copper or 1750 kcmil aluminum, the grounded conductor must not be smaller than 12½% of the area of the largest conductor. Where the service-entrance phase conductors are paralleled, the size of the grounded conductor is to be based on the equivalent area for parallel conductors as indicated in this section.

Exception No. 3 applies to this sizing rule and says that the neutral conductor of DC generators that must carry ground-fault currents is not to be smaller than the minimum required size of the largest phase conductor.

Sec. 445-6 — Protection of live parts

Live parts of generators operated at more than 50V to ground must not be exposed to accidental contact where accessible to unqualified persons.

Recall that the definition of accessible as applied to equipment in Article 100 is:

> not guarded by locked doors, elevation, or other effective means.

Sec. 445-7 — Guards for attendants

Where necessary for the safety of attendants, the requirements of **Sec. 430-133** apply. There is says that where live parts operating at over 150V to ground are guarded against accidental contact only by location as specified in **Sec. 430-132**, and where adjustment or other attendance may be necessary during the operation of the apparatus, suitable insulating mats or platforms must be provided so that the attendant cannot readily touch live parts unless standing on the mats or platforms. An FPN acts as a reminder of the need for adequate working space per **Secs. 110-16** and **-34**.

The locations permitted by Sec. 430-132 to substitute for providing guarding for equipment are:

- in a room or enclosure accessible only to qualified persons;
- on a suitable balcony, gallery, or platform, elevated and arranged to exclude unqualified persons; or
- by mounting at an elevation 8 ft (2.44 m) or more above the floor.

Sec. 445-8 — Bushings

Where wires pass through an opening in an enclosure, conduit box, or barrier, a bushing must be used to protect the conductors from the edges of an opening having sharp edges. The bushing must have smooth, well-rounded surfaces where it may be in contact with the conductors. If used where oil, grease, or other contaminants may be present, the bushing must be made of a material that is not affected by them.

ARTICLE 250
GROUNDING

Auxiliary power sources used as emergency systems, legally required standby systems, or optional standby systems are required to be fixed in place. There, however is no similar requirement that the system be permanently fixed for the power source to qualify as being an "interconnected electric power production source." Auxiliary power systems, thus, also can include those that can be moved from one place to another within the facility in order to supplement or substitute for normal power. Grounding of both these types of units are covered by the NEC.

Sec. 250-5(d) — Separately derived systems

This section defines what is and what is not a separately derived system. The distinction is crucial to assuring proper grounding of the onsite auxiliary power source.

A premises wiring system whose power is derived from a generator or converter winding and has no direct electrical connection, including a solidly connected grounded circuit conductor, to supply conductors originating in another system, is required to be grounded as specified in **Sec. 250-26**.

Per **Sec. 250-5(b)**, an AC system operating at 50 to 1000V is required to be grounded where:

- it can be so grounded that the maximum voltage to ground on the ungrounded conductors does not exceed 150V; or
- the system is 3-phase 4-wire wye connected in which the neutral is used as a circuit conductor; or

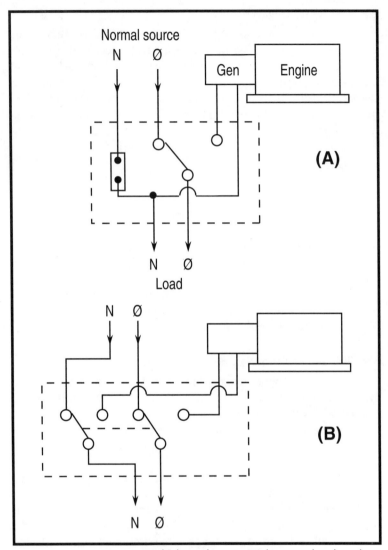

■ **Fig. 6.3.** In a non-separately derived system **(A)** the normal and auxiliary source neutrals are connected together; in the separately derived system **(B)** the neutrals are separated by the pole of the transfer switch. (Only one phase conductor is shown for simplicity.)

- the system is 3-phase 4-wire delta connected in which the midpoint of one phase winding is used as a circuit conductor.

FPN No. 1 then notes that an onsite AC generator is *not* a separately derived system if its neutral is solidly interconnected to a service-supplied system neutral. FPN No. 2 further states that systems not separately derived, thus not required to be grounded as specified in Sec. 250-26, must have minimum sized conductors per Sec. 445-5 to carry fault current.

The difference between the way the grounded (neutral) conductor is connected to the load (as seen in **Fig. 6.3**) is a determining factor in deciding whether the system is a separately derived system or not. For the requirements of grounding a separately derived AC system, see the "Understanding NEC Grounding" book referred to earlier.

Sec. 250-6 — Portable and vehicle-mounted generators

Those systems that are not permanently fixed cannot qualify as emergency, legally required, or optional systems because they may not be at the required location when needed. There are, however, sets of rules on grounding applicable to these units that must be adhered to.

(a) Portable generators. The frame of a portable generator is not required to be grounded and is permitted to serve as the grounding electrode for a system supplied by the generator if the following conditions are met:

- the generator supplies only equipment mounted on the generator or cord-and-plug-connected equipment through receptacles mounted on the generator, or both; and
- the noncurrent-carrying metal parts of equipment and the equipment grounding conductor terminals of the receptacles are bonded to the generator frame.

(b) Vehicle-mounted generators. The frame of a vehicle is permitted to serve as the grounding electrode for a system supplied by a generator located on the vehicle if the following conditions are met:

- the frame of the generator is bonded to the vehicle frame;

- the generator supplies only equipment located on the vehicle or cord-and-plug-connected equipment through receptacles mounted on the vehicle, or both equipment mounted on the vehicle and cord-and-plug-connected equipment through receptacles mounted on the vehicle or on the generator;
- the noncurrent-carrying metal parts of equipment and the equipment grounding conductor terminals of the receptacles are bonded to the generator frame; and
- the system complies with all other provisions of Article 250.

(c) Neutral conductor bonding. A neutral conductor must be bonded to the generator frame where the generator is a component of a separately derived system. The bonding of any conductor other than a neutral within the generator to its frame is not required.

ARTICLE 690 — SOLAR PHOTOVOLTAIC SYSTEMS

7.

In discussing onsite auxiliary power sources, it must be remembered that solar photovolatic systems also are being used for this purpose, and their application for this purpose probably will increase in the future. They can produce power either on a standalone basis or as part of an interactive electric power production system with other sources.

Solar photovoltaic systems covered by this article can have AC inverter or DC outputs, and can be applied with or without batteries that provide electrical energy storage capacity.

ARTICLE 690, PART A
SOLAR PHOTOVOLTAIC SYSTEMS — GENERAL

Sec. 690-1 — Scope

A typical arrangement of a solar photovoltaic system is shown in **Fig. 7.1** on the next page. The item marked "A" is the disconnecting means required by rules discussed later in this chapter. The items marked "B" are equipment permitted to be on the photovoltaic power source side of the photovoltaic power source disconnecting means per **Sec. 690-14**. Note that grounded conductors are not shown and, thus, it is *not* the intent of Fig. 7.1 to require a disconnect in grounded circuit conductors.

Sec. 690-2 — Definitions

The following definitions of various terms used in this article to describe solar photovoltaic systems and their various components.

Array: A mechanically integrated assembly of modules or panels with a support structure and foundation, tracking, thermal control, and other components, as required, to form a DC power producing unit.

Blocking diode: A diode used to block reverse flow of current into a photovoltaic source circuit.

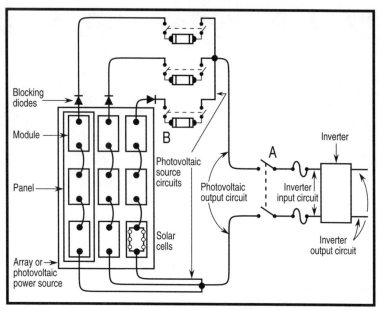

■ **Fig. 7.1.** Typical solar photovoltaic system. Adapted from NEC Diagram 690-1.

Interactive system: A solar photovoltaic system that operates in parallel with another electric power production source connected to the same load. A battery or other energy storage device that is a subsystem of a solar photovoltaic system is not considered to be another electric power production source.

Inverter: Equipment that is used to change voltage level or waveform, or both, or electrical energy. Commonly, an inverter (also known as a power conditioning unit (PCU) or power conversion system (PCS) is a device that changes DC input to an AC output. Inverters in stand-alone systems might also include battery chargers that take AC from an auxiliary source, such as a generator, and rectify it into DC for charging batteries.

Inverter input circuit: Conductors between the inverter and the battery in stand-alone systems, or the conductors between the inverter and the PV ouput circuits for grid-connected systems.

Inverter output circuit: Conductors between the inverter and an AC load center for stand-alone systems, or the conductors between the inverter and the service equipment or other electric power production source (such as a utility) for grid-connected systems.

Module: The smallest complete, environmentally protected assembly of solar cells, optics, and other components, exclusive of tracking, designed to generate DC power under sunlight.

Panel: A collection of modules mechanically fastened together, wired, and designed to provide a field-installable unit.

Photovolatic output circuit: Circuit conductors between the photovoltaic source circuit(s) and the power conditioning unit or DC utilization equipment.

Photovoltaic power source: An array of aggregate of arrays that generates DC power at system voltage and current.

Photovoltaic source circuit: Conductors between modules and from modules to the common connection point(s) of the DC system.

Solar cell: The basic photovoltaic device that generates electricity when exposed to light.

Solar photovoltaic system: The total components and subsystems that, in combination, convert solar energy into electrical energy suitable for connection to a utilization load.

Stand-alone system: A solar photovoltaic system that supplies power independently, but may receive control power from another electric power production source.

Sec. 690-3 — Other articles

In the arrangement of the Code articles, those in Chapters 1 through 5 apply generally to all electrical circuits and equipment. Chapter 6 and others that follow contain rules of articles involving specific types of installations. Where different from the provisions in Chapters 1-5, it is the rules in the specific articles that apply. For instance, the regulations in Article 690 on solar photovoltaic systems supersede any regulation in Chapters 1-5 with which they conflict. Note, however, all other applicable rules in Chapters 1-5 that *do not conflict* must be adhered when designing and installing such a system.

Solar photovoltaic systems operating as interconnected power production sources must be installed in accordance with the provisions of Article 705 covered in Chapter 5 of this book.

Sec. 690-4 — Installation

A solar photovoltaic system is permitted to supply electricity to a building or structure. This source can be in addition to power supplied by other electric services to the same building or structure. A solar photovoltaic system, thus, can be used as an alternate supply.

Note that a photovoltaic system is not listed as an acceptable alternate source of supply for an emergency or legally required standby system. The reason is that a solar photovoltaic system is not considered to be a reliable enough source to permit such applications. Optional standby systems have no such restrictions, nor do interconnected electric power production sources. As a matter of fact, Sec. 705-3 lists Article 690 on solar photovoltaic systems as a reference.

Photovoltaic source and output circuits are not permitted to be contained in the same raceway, cable tray, cable, box, or fitting as feeders of branch circuits of other systems. An exception to this rule is made if the different systems are separated by a partition, or are connected together (as shown in **Fig. 7.2**).

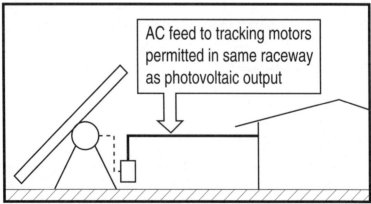

■ **Fig. 7.2.** In some instances, the photovoltaic system conductors and other systems are permitted within the same raceway.

Connections must be so arranged that removal of a module or panel from a photovoltaic source circuit does not interrupt the grounded conductor to other photovoltaic source circuits.

Inverters or motor generators used as part of a solar photovoltaic system must be identified specifically for such use.

Sec. 690-5 — Ground fault detection arrangement

Roof-mounted photovoltaic arrays of dwellings must be provided with ground-fault protection to reduce fire hazard. The GFP circuit must be capable of detecting a ground fault, interrupting the fault path, and disabling the array.

For more information on ground-fault protection, see *EC&M's* book entitled *Practical Guide to Ground Fault Protection*.

ARTICLE 690, PART B
SOLAR PHOTOVOLTAIC SYSTEMS — CIRCUIT REQUIREMENTS
Sec. 690-7 — Maximum voltage

The voltage of a photovoltaic power source, and its DC circuits, is defined as being its rated open-circuit voltage. For 3-wire installations, including 2-wire circuits connected to 3-wire systems, the system voltage is considered to be the highest rated voltage between any two conductors.

The DC voltage of utilization circuits must conform with **Sec. 210-6.** This reference sets the following limitations on branch circuit voltage.

- In dwelling units and guest rooms of hotels, motels, etc., the voltage must not exceed 120V between conductors, and is limited to supplying lighting fixtures and cord-and-plug connected loads rated 1440VA or 1/4 hp, or less.

- In other buildings or structures, circuits not exceeding 120V can supply lampholders applied within their voltage rating, auxiliary equipment of electric-discharge lamps, and cord-and-plug connected or permanently connected utilization equipment.

- Circuits over 120V but not exceeding 277V to ground can supply listed lighting fixtures, auxiliary equipment of electric discharge lamps, and cord-and-plug permanently connected utilization equipment.

• Circuits exceeding 277V to ground but not exceeding 600V between conductors are permitted to feed auxiliary equipment of electric-discharge lamps permanently mounted not less than 22 ft above grade on poles for outdoor illumination, or 18 ft in other structures; and to supply cord-and-plug connected permanently connected utilization equipment.

Photovoltaic source and output circuits not containing lampholders, fixtures, or receptacles are permitted up to 600V. For other than one- or two-family dwelling units, systems over 600V are permitted. Their wiring must comply with the provisions of **Article 710**.

In one- or two-family dwellings, energized live parts of photovoltaic source or output circuits over 150V to ground must not be accessible to other than qualified persons.

Sec. 690-8 — Circuit sizing and current

The ampacity of conductors and rating or setting of overcurrent devices in a circuit of a solar photovoltaic system must not be less than 125% of the current in the circuit computed as follows.

Photovoltaic source circuits: The sum of parallel module short-circuit current ratings.

Photovoltaic output circuits: The photovoltaic power source short-circuit current rating.

Inverter output circuits: The inverter or power conditioning unit output current rating. An exception is made for cases where the output circuit does not include an overcurrent device. In this instance, if there is no external power source that can backfeed through the conductors, the current rating is allowed to be the short-circuit current.

Stand-alone inverter input circuit: The input current rating when the inverter is producing rated power at the lowest input voltage.

An exception to the 125% rule is made for circuits containing an assembly together with its overcurrent devices that is listed for continuous operation at 100% of its rating.

The rating or setting of overcurrent devices is permitted in accordance with **Sec. 240-3(b)** and **(c)**. The requirements there are as follows.

- The next higher standard overcurrent device rating (above the ampacity of the conductors being protected) is permitted, but only if all the following conditions are met: the conductors being protected are not part of a multioutlet branch circuit supplying receptacles for cord-and-plug portable loads; the ampacity of the conductors does not correspond with the standard ampere rating of a fuse or circuit breaker without overload trip adjustments above its rating (but is allowed to have other trip or rating adjustments); and if the next higher rating does not exceed 800A.

- Where the overcurrent device is rated over 800A, the ampacity of the conductors it protects must be equal to, or greater than, the standard ampere rating of the overcurrent device. In **Sec. 240-6**, these are given as 1000, 1200, 1600, 2000, 2500, 3000, 4000, 5000, and 6000A.

For a photovoltaic power source having multiple output circuit voltages and employing a common return conductor, the ampacity of the common-return conductor, as seen in **Fig. 7.3**, must not be less than the sum of the ampere rating of the overcurrent devices of the individual output circuits.

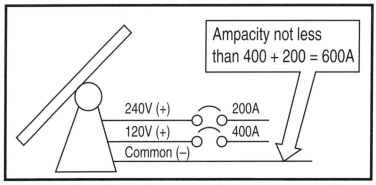

■ **Fig. 7.3.** This is the way the ampacity of a common return conductor is calculated for systems having multiple DC voltages and employing a common return conductor.

Sec. 690-9 — Overcurrent protection

Photovoltaic source and output circuits, power conditioning unit output circuit, and storage battery circuit conductors and equipment must be protected in accordance with the requirements of **Article 240**.

In general, the important requirements are of the sizing of the protective devices. These rules were covered in Sec. 690-8.

Circuits connected to more than one electrical source must have overcurrent devices so located that they provide overcurrent protection from all sources. The concern here is preventing high-current backfeeds from other sources. A thorough analysis of the circuit protection is needed.

Overcurrent protection for a transformer with a source on each side must be provided by considering first one side of the transformer, then the other side, as the primary. For specifics on overcurrent protection of transformers, refer to *EC&M's* book entitled *Understanding NE Code Rules on Transformers*.

An exception to the transformer protection rule is made for a power transformer with a current rating of not less than the short-circuit output rating of the power conditioning unit on the side connected toward the photovoltaic power source. In such a case, it is permitted to omit the overcurrent protection from that source.

Branch-circuit or supplementary type overcurrent devices are permitted to provide overcurrent protection in photovoltaic source circuits. The overcurrent devices must be accessible, but are not required to be readily accessible.

Overcurrent devices (fuses or circuit breakers) used in any DC portion of a photovoltaic power system must have the appropriate DC voltage, current, and interrupt ratings.

ARTICLE 690, PART C
SOLAR PHOTOVOLTAIC SYSTEMS — DISCONNECTING MEANS
Sec. 690-13 — All conductors

Means must be provided to disconnect all current-carrying conductors of a photovoltaic power source from all other conductors in a building or other structure. This is the general rule, but there is a major qualification.

As shown in **Fig. 7.4,** an exception says that the disconnect means must not interrupt the grounded conductor except if the disconnect is part of a ground-fault protection system required by Sec. 690-5. Such a GFP system is required only on rooftop arrays of dwelling units.

The reasoning behind this rule is that in photovoltaic arrays, only one grounding connection can be made between the grounded conductor and the grounding electrode conductor in order to avoid objectionable currents from flowing in the equipment grounding system in parallel with the grounded conductor. If this grounding connection is on the load side of the disconnecting means (as shown) and the switch is opened, the photovoltaic array floats at an unknown voltage to ground. This is a potential hazard, particularly in large arrays. They can be subjected to induced voltage spikes that will cause hidden damage to insulation. On the other hand, if the grounding connection is on the line side of the disconnect, it will be impossible to stop a ground fault in progress using a device that interrupts the grounding conductor.

An FPN, however, explains that a separate mechanical means of opening the grounded conductor, such as a removable link or a terminal, is permitted to allow for maintenance or troubleshooting by qualified personnel.

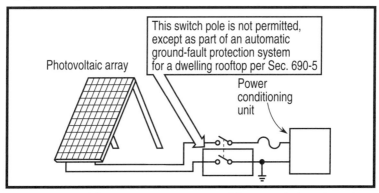

■ **Fig. 7.4.** Prohibition against interruption of the grounded conductor by a disconnecting device.

Sec. 690-14 — Additional provisions

The provisions of **Article 230, Part F** on service equipment disconnecting means also apply to photovoltaic installations. These rules are extensive and beyond the scope of this book. For a detailed look at the rules, see *EC&M's* book entitled, *Understanding NE Code Rules on Services*.

Exceptions are made to Article 230 requirements for service disconnect means when applied to photovoltaic installations. They provide that disconnects used in photovoltaic systems are not required to be suitable as service equipment and are to be rated as required in Article 690.

Another exception allow equipment such as isolating switches, overcurrent devices, and blocking diodes on the power source side of the disconnecting means of a photovoltaic system. This can be best understood by again referring back to Fig. 7.1. If derived from a utility or onsite generator, no equipment (except metering devices) would be allowed on the power source side of the disconnect switch A.

Sec. 690-15 — Disconnection of photovoltaic equipment

Means must be provided to disconnect equipment, such as the inverter, batteries, chargers, and the like, from all ungrounded conductors of all sources. If the equipment is energized from more than one source, the disconnecting means must be grouped and identified.

Sec. 690-16 — Fuses

Disconnecting means must be provided to disconnect a fuse from all sources of supply if the fuse is energized from both directions and is accessible to other than qualified persons.

In the case of the fuses marked "B" in Fig. 7.1, they must be grouped and identified per Sec. 690-15, and disconnected on both sides because they are tied together on their "output" side and therefore could have power available on that side even if the switch on their "input" side is open.

Such a fuse in a photovoltaic source circuit most be capable of being disconnected independently of fuses in other photovoltaic source circuits. In other words, one fuse in the photovoltaic output circuit conductors cannot be substituted for the individual B fuses.

Sec. 690-17 — Switch or circuit breaker

The disconnecting means for ungrounded conductors must consist of a manually operable switch or circuit breaker meeting the following requirements. They are to:
- be located where readily accessible;
- be externally operable without exposing the operator to contact with live parts;
- indicate clearly whether in the open or closed position; and
- have a interrupting rating sufficient for the nominal circuit voltage and the current available at the line terminals of the equipment.

An exception says that a disconnecting means located on the DC side is permitted to have an interrupting rating less than the current-carrying rating when interlocked so that the DC switch cannot be opened under load.

Where it is possible that all terminals of the disconnecting means could be energized while the switch is in the open position, a warning sign must be mounted on, or adjacent to, the disconnecting means. The sign must clearly state: WARNING — ELECTRIC SHOCK — DO NOT TOUCH — TERMINALS ENERGIZED IN OPEN POSITION.

Sec. 690-18 — Disablement of an array

Means must be provided to disable an array or portions of an array. An FPN explains that persons installing, replacing, or servicing components of an array may receive a shock if the modules are exposed to the sun and, thus, energized.

ARTICLE 690, PART D
SOLAR PHOTOVOLTAIC SYSTEMS — WIRING METHODS

Sec. 690-31 — Methods permitted

Wiring systems. All raceway and cable wiring methods included in the NEC, and other wiring systems and fittings specifically intended and identified for use on photovoltaic arrays, are permitted.

Where wiring devices with integral enclosures are used, sufficient length of cable must be provided to facilitate replacement.

Single conductor cable, Types SE, UF, and USE, are permitted in photovoltaic source circuits where installed in the same manner as a Type UF multiconductor cable in accordance with **Article 339**. Where exposed to the direct rays or the sun, Type UF cable identified as sunlight-resistant or Type USE cable must be used.

The reference to Article 339 is vague since Sec. 339-3 is the only one that seems to apply, and that section deals with permitted uses rather than installation methods. The following items are listed there.

• Multiconductor UF cable is permitted for use as an underground feeder or branch circuit, including direct burial in the earth, if provided with overcurrent protection at their ampacity.

• For single conductors UF, all conductors including the neutral must be run together in the same trench or raceway. An exception, however, says that this provision does not apply to solar photovoltaic systems.

• Type UF cable is also permitted for interior wiring in wet, dry, or corrosive locations. Where installed as nonmetallic-sheathed cable, it must be of the multiconductor type. Here again, an exception permits the use of single-conductor UF cable in solar photovoltaic systems.

• For underground requirements, see **Sec. 300-5**.

An FPN refers to the FPN following **Sec. 310-13** for information on the use of insulated cables for photovoltaic source circuits. It says there that when thermoplastic-insulated conductors are used on DC circuits in wet locations, the ends of the insulation at terminals might deteriorate because of a reaction between conductor and insulation that is termed "electroendosmosis."

Flexible cords and cables, where used to connect the moving parts of tracking photovoltaic modules, must comply with **Article 400** and must be: of a type identified as hard-service cord or portable-power cable; suitable for extra-hard usage; and listed for outdoor use and water and sunlight resistance. Allowable ampacities must be in accordance with **Sec. 400-5**. For ambient temperatures

Ambient Temp. °C	Temperature Rating of Conductor			
	60°C	75°C	90°C	105°C
30	1.00	1.00	1.00	1.00
31-35	0.91	0.94	0.96	0.97
36-40	0.82	0.88	0.91	0.93
41-45	0.71	0.82	0.87	0.89
46-50	0.58	0.75	0.82	0.86
51-55	0.41	0.67	0.76	0.82
56-60	–	0.58	0.71	0.77
61-70	–	0.33	0.58	0.68
71-80	–	–	0.41	0.58

■ **Fig. 7.5.** Correction factors for flexible cord or cables used with photovoltaic equipment when the ambient temperature exceeds 30°C. Adapted from NEC Table 690-31(c).

exceeding 30°C, the ampacities are to be derated by the factors given in **Fig. 7.5**.

Small-conductor cables. Listed sunlight and moisture resistant single-conductor cables in sizes No. 16 and No. 18 are permitted for module interconnections where such cables meet the ampacity requirements discussed in Sec. 690-8. **Sec. 310-15** is to be used to determine the cable ampacity and temperature derating factors.

Sec. 690-32 — Component interconnections

Fittings and connectors intended to be concealed at the time of onsite assembly, where listed for such use, are permitted for onsite interconnection of modules or other array components. Such fittings and connectors must be equal to the wiring method employed in insulation, temperature rise, and fault-current withstand, and must be capable of resisting the effects of the environment in which they are used.

Sec. 690-33 — Connectors

Permitted connectors must be:
- polarized and have a configuration that is noninterchangeable with receptacles in other electrical systems on the premises;
- constructed and installed so as to guard against inadvertent

contact with live parts by persons;
- of the latching or locking type;
- configured so that the grounding member is the first to make and last to break contact with the mating connector; and
- capable of interrupting the circuit current without hazard to the operator.

The key consideration when a solar photovoltaic system is installed is that the receptacle configuration be different from any others on the premises. For instance, if only straight pin and locking type receptacles are employed, pin-and-sleeve connectors, or possibly a unique configuration of blades can be used.

Sec. 690-34 — Access to boxes

Junction, pull, and outlet boxes located behind modules or panels must be installed so that the wiring contained in them can be rendered accessible directly or by displacement of modules or panels secured by removable fasteners and connected by a flexible wiring system.

ARTICLE 690, PART E
SOLAR PHOTOVOLTAIC SOURCES — GROUNDING
Sec. 690-41 — System grounding

For a photovoltaic power source, one conductor of a 2-wire system rated over 50V and a neutral conductor of a 3-wire system must be solidly grounded.

An exception says that other methods that accomplish equivalent system protection and that utilize equipment listed and identified for the use are also permitted. An FPN refers to others following Sec. 250-1. There the purpose of system and circuit grounding is defined as limiting voltage due to lightning, surges, etc., and to stabilize voltage to ground.

Sec. 690-42 — Point of system grounding connection

The DC circuit grounding connection must be made at only one single point on the photovoltaic ouput circuit. An FPN, further, notes that locating the grounding connection point as close as practicable to the photovoltaic source will better protect the system from voltage surges due to lightning.

Sec. 690-43 — Equipment grounding

Exposed noncurrent-carrying metal parts of module frames, equipment, and conductor enclosures must be grounded regardless of voltage. This applies even where the photovoltaic output circuits are ungrounded. There are concerns about high short-circuit currents from battery banks connected to this equipment as well as lightning-induced surge currents.

Sec. 690-45 — Size of equipment grounding conductor

The equipment grounding conductor must be no smaller than

Rating or Setting of Automatic Overcurrent Device in Circuit Ahead of Equipment, Conduit, etc., Not Exceeding (Amperes)	Size	
	Copper Wire No.	Aluminum or Copper-Clad Aluminum Wire No.
15	14	12
20	12	10
30	10	8
40	10	8
60	10	8
100	8	6
200	6	4
300	4	2
400	3	1
500	2	1/0
600	1	2/0
800	1/0	3/0
1000	2/0	4/0
1200	3/0	250 kcmil
1600	4.0	350 kcmil
2000	250 kcmil	400 kcmil
2500	350 kcmil	600 kcmil
3000	400 kcmil	600 kcmil

■ **Fig. 7.6.** Minimum size of equipment grounding conductors. Adapted from NEC Table 250-95.

the required size of the circuit conductors in systems where the available photovoltaic power source short-circuit current is *less* than twice the current rating of the overcurrent device.

In other systems, the equipment grounding conductor must be sized in accordance with Sec. 250-95. The sizing of the conductor given there is shown in the table of **Fig.** 7.6 on the previous page.

Note that the purpose of equipment grounding is to supply a low impedance path for fault current that will facilitate the operation of overcurrent devices under ground fault conditions. Therefore, if the photovoltaic source short-circuit current is relatively low, a lower impedance path (larger conductor) is required to assure operation of the protective equipment. Where it is relatively high, a reduced size of conductor is acceptable.

Sec. 690-47 — Grounding electrode system

A grounding electrode system must be provided in acccordance with **Secs.** 250-81 through -86. These sections make up Part H of Article 250, which describes grounding electrode systems permitted by the Code. For a detailed discussion of the subject, see *EC&M's* book entitled, *Understanding NE Code Rules on Grounding & Bonding.*

ARTICLE 690, PART F
SOLAR PHOTOVOLTAIC SOURCES — MARKING
Sec. 690-51 — Modules

Modules must be marked with:
- identification of the polarity of terminals or leads;
- maximum overcurrent device rating for module protection;
- rated open-circuit voltage;
- operating voltage;
- maximum permissible system voltage;
- operating current;
- short-circuit current; and
- maximum power.

Sec. 690-52 — Photovoltaic power sources

A marking is to be provided *by the installer* at an accessible location at the disconnecting means for the photovoltaic power source, specifying the photovoltaic power source rated:

- operating current;
- operating voltage;
- open-circuit voltage; and
- short-circuit current.

An FPN comments that reflecting systems used for focusing sunlight and enhancement may result in increased levels of output current and power.

ARTICLE 690, PART G
SOLAR PHOTOVOLTAIC SOURCES — CONNECTION TO OTHER SOURCES
Sec. 690-61 — Loss of system voltage

Power output from a power conditioning unit in a solar photovoltaic system is sometimes interactive with another electric system, such as an electric utility supply to a facility. This is permitted in **Article 705** for interconnected electric power production sources. When this is the case, the power conditioning unit output must be automatically disconnected from all ungrounded conductors in the other electric system (or systems) upon loss of voltage in the other electric system. The two systems must not be reconnected until the other system's voltage is restored.

It should be noted, however, that the normally interactive solar photovoltaic system is permitted to continue to operate as a stand-alone system to supply premises wiring even though it has been disconnected from the other power source.

Sec. 690-62 — Ampacity of neutral conductor

A single-phase, 2-wire power conditioning unit output can be connected to the neutral and one ungrounded conductor (only) of a 3-wire system or 3-phase, 4-wire wye-connected system as shown in **Fig. 7.7** on the next page. The maximum load connected between

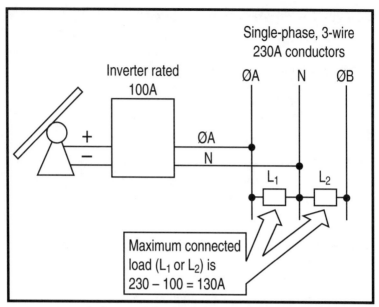

■ **Fig. 7.7.** Calculation of the load that can be carried between any conductor and the neutral when a 2-wire output of a power conditioning unit is tied into a single-phase, 3-wire system (or a 3-phase, 4-wire system).

the neutral and any one ungrounded conductor, plus the power conditioning unit output ampere rating, must not exceed the ampacity of the neutral conductor.

Sec. 690-63 — Unbalanced interconnections

The output of a single-phase photovoltaic power conditioning unit is forbidden to be connected to a 3-phase, 3- or 4-wire electrical service derived directly from a delta-connected transformer.

On the other hand, a 3-phase photovoltaic power conditioning unit can be connected to such systems, but must be automatically disconnected from all ungrounded conductors of the interconnected system when one of the phases opens in either source.

An exception applicable to both the single-phase and 3-phase photovoltaic power conditioning units is made where the interconnected system is designed so that significant unbalanced voltages will not result.

Sec. 690-64 — Point of connection

In general, the output of a photovoltaic power conditioning unit where connected to an AC electrical source, or the photovoltaic output circuit where interactive with a DC electric source, must be connected to the supply side of the service disconnecting means (permitted in **Sec. 230-82, Exception No. 6**).

An option is to connect the photovoltaic output to the load side of the service disconnecting means of the other source. For this to be permitted, however, *all* the following conditions have to be met.

- Each source interconnection must be made at a dedicated circuit breaker or fusible disconnecting means.
- The sum of the ampere ratings of overcurrent devices in circuits supplying power to a busbar or conductor must not exceed the rating of the busbar or conductor. In a dwelling unit, however, the sum of the ampere rating of the overcurrent devices must not exceed 120% of the busbar or conductor rating.
- The interconnection point must be on the line side of all ground-fault protection equipment. A connection, however, is permitted to be made to the load side of ground-fault protection, provided there is GFP for equipment from all ground-fault current sources (see Fig. 5.4 in Chapter 5).
- Equipment containing overcurrent devices in circuits supplying power to a busbar or conductor must be marked to indicate the presence of all sources. An exception is made for equipment with power supplied from a single point of connection.
- Equipment such as circuit breakers, if backfed, must be identified for such operation.

ARTICLE 690, PART H
SOLAR PHOTOVOLTAIC SYSTEMS — STORAGE BATTERIES

Sec. 690-71 — Installation

Storage batteries in a solar photovoltaic system must be installed in accordance with the provisions of **Article 480**.

The principal requirements given there deal with necessity for providing additional insulating support. For batteries in a string operating at not over 250V, the following is required.

Vented lead-acid batteries. None required for cells and multicompartment units with covers sealed to jars of nonconductive, heat-resistant material.

Vented alkaline-type batteries. None required for cells with covers sealed to jars of nonconductive, heat resistant material. If, however, the jars are of conductive material, they must be installed in trays of nonconductive material, with no more than 20 cells (24V nominal) in the series circuit in any one tray.

Rubber jars. None required where the total nominal voltage of all cells in series does not exceed 150V. Where higher, batteries must be sectionalized into groups of 150V or less, and each group must be on a separate tray or rack.

Sealed cells or batteries. None required if constructed of nonconductive, heat-resistant material. Those constructed of conducting material must have insulating support if a voltage is present between the container and ground.

The same requirements apply to batteries that are part of a string operating at over 250V. In addition, the cells must be installed in groups having a total nominal voltage of not over 250V. Insulation (which can be air) must be provided between groups and have a minimum separation between live battery parts of opposite polarity or 2 inches (50.8 mm) for battery voltages not exceeding 600V.

Racks must be substantial and made of metal resistant to deterioration by the electrolyte, fiberglass, or other suitable nonconductive material.

Provisions for ventilation to prevent an accumulation of an explosive mixture, and for the guarding of live parts, must also be made.

Storage batteries for dwellings must have the cells connected so as to operate at less than 50V. An exception to this rule applies where live parts are not accessible during routine battery maintenance. In such cases, battery system voltage in accordance with Sec. 690-7 is permitted.

Live parts of battery systems for dwellings must be guarded to prevent accidental contact by persons or objects, regardless of voltage or battery type.

An FPN notes that batteries in solar photovoltaic systems are subject to extensive charge-discharge cycles and typically require frequent maintenance, such as checking electrolyte and cleaning connections.

A listed current-limiting overcurrent device must be installed in each circuit adjacent to the batteries where the available short-circuit current from a battery or battery bank exceeds the ratings of other equipment in that circuit. If current-limiting fuses are used for this purpose, they must comply with Sec. 690-16. The requirement there is for disconnectability from any source of supply.

Sec. 690-72 — State of charge

Equipment must be provided to control the state of charge of the battery. All adjusting means for control of the state of charge must be accessible only to qualified persons. An exception is made where the design of the photovoltaic power source is matched to the voltage rating and charge current requirement for the interconnected battery cells.

Sec. 690-73 — Grounding

The interconnected battery cells are to be considered grounded where the photovoltaic power source is installed in accordance with the exception to Sec. 690-41, which was discussed earlier in this chapter.

Sec. 690-74 — Battery interconnections

Flexible cables identified in **Article 400**, in sizes No. 2/0 and larger, are permitted within the battery enclosure from battery terminals to nearby junction box where they must be connected to an approved wiring method. Flexible battery cables are also permitted between batteries and cells within the battery enclosure. Such cables must be listed for hard service use and identified as being acid and moisture resistant.

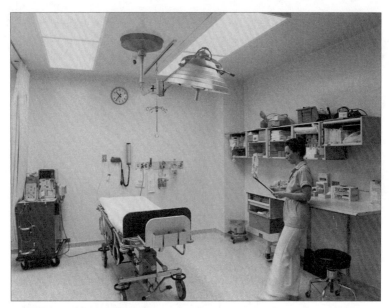

Health-care facilities are major users of auxiliary power sources. Article 517 contains rules for deciding when they must be installed.

ARTICLE 517 — HEALTH CARE FACILITIES

There are few places where the need for onsite power is critical as in health-care facilities. Lives, as well as medical records and other items are at risk when the prime source of power is lost. Thus, the requirements for alternate power sources are addressed in this article. Only those sections that directly address the requirements for emergency or standby systems are covered in this chapter.

ARTICLE 517, PART A
HEALTH CARE FACILITIES — GENERAL
Sec. 517-3 — Definitions

Among the items whose meaning are defined are the following that relate to the subject of onsite power.

Alternate power source: One or more generator sets, or battery systems where permitted, intended to provide power during the interruption of the normal electrical services or the public utility electrical service intended to provide power during interruption of service normally provided by the generating facilities on the premises.

Ambulatory Health Care Center: A building or part thereof used to provide services or treatment for four or more patients at the same time and meeting either (1) or (2) below.

(1) Those facilities that provide, on an outpatient basis, treatment for patients that would render them incapable of taking action for self-preservation under emergency conditions without assistance from others, such as hemodialysis units or freestanding emergency medical units.

(2) Those facilities that provide, on an outpatient basis, surgical treatment requiring general anesthesia.

Critical branch: A subsystem of the emergency system consisting of feeders and branch circuits supplying energy to task illumination, special power circuits, and selected receptacles serving areas and functions related to patient care, and which are connected to alternate power sources by one or more transfer switches during interruption of the normal power source.

Electrical life-support equipment: Electrically powered equipment whose continuous operation is necessary to maintain a patient's life.

Emergency system: A system of feeders and branch circuits meeting the requirements of Article 700, and intended to supply alternate power to a limited number of prescribed functions vital to the protection of life and patient safety, with automatic restoration of electrical power within 10 seconds of power interruption.

Equipment system: A system of feeders and branch circuits arranged for delayed, automatic or manual connection to the alternate power source and which serves primarily 3-phase power equipment.

Essential electrical system: A system comprised of alternate sources of power and all connected distribution systems and ancillary equipment, designed to assure continuity of electrical power to designated areas and functions of a health care facility during disruption of normal power sources, and also designed to minimize disruption within the internal wiring system.

Life safety branch: A subsystem of the emergency system consisting of feeders and branch circuits, meeting the requirements of Article 700 and intended to provide adequate power needs to ensure safety to patients and personnel, and which are automatically connected to alternate power sources during interruption of the normal power source.

Limited care facility: A building or part thereof used on a 24-hr basis for the housing of four or more persons who are incapable of self preservation because of age, physical limitation due to accident or illness, or mental limitations, such as mental retardation/developmental disability, metal illness, or chemical dependency.

Nursing home: A building of part thereof used for the lodging, boarding and nursing care, on a 24-hr basis, or four or more persons who, because of mental or physical incapacity, may be unable to provide for their own needs and safety without the assistance of another person. Nursing home, wherever used in the Code, includes nursing and convalescent homes, skilled nursing

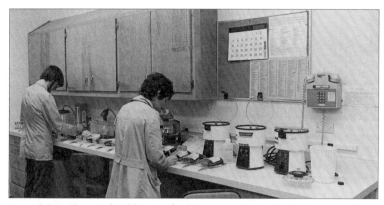

Critical branches in health care facilities can also include those supplying power to blood laboratories and others that are essential for patient care.

facilities, intermediate care facilities, and infirmaries of homes for the aged.

Task illumination: Provision for the minimum lighting required to carry out necessary tasks in the described areas, including safe access to supplies and equipment, and access to exits.

ARTICLE 517, PART B
HEALTH CARE FACILITIES — WIRING AND PROTECTION
Sec. 517-10 — Applicability

Part B applies to all health care facilities. Several exceptions, however, list those where the rules of Part B *do not* apply.

Exception No. 1: Business offices, corridors, waiting rooms, and the like in clinics, medical and dental offices, and outpatient facilities.

Exception No. 2: Patient sleeping areas in nursing homes and limited care facilities wired in accordance with NEC Chapters 1 through 4. This exception does not apply to areas where patients are intended to be placed on life support ystems or subjected to invasive porcedures and connected to line-operated electromedical devices.

An FPN refers to NFPA 101, *Life Safety Code*. A further

discussion of this standard is included in this chapter.

Sec. 517-17 — Ground-fault protection

(a) Feeders. This subsection requires that where ground-fault protection is required (1000A at 150 to 600V), an additional step of ground-fault protection must be provided for the next level of feeder disconnecting means downstream toward the load (**Fig. 8.1A**).

As shown in (**B**), an FPN says this second level of ground-fault protection *is not to be installed*:

• on the load side of an essential electrical system (see defini-

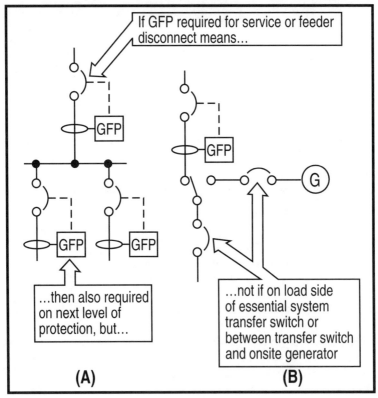

■ **Fig. 8.1.** Requirement for ground-fault protection of service or feeder disconnect means in health care facilities.

tion) transfer switch;
- between the onsite generating unit(s) required in Sec. 517-35(b) and essential electrical system transfer switches; and
- on electrical systems that are not solidly grounded wye systems with greater than 150V to ground, but not exceeding 600V phase-to-phase.

Sec. 517-18 — General care areas

(a) Patient bed location. Each patient bed location is required to have at least one branch circuit fed from the emergency system. This is in addition to the one or more branch circuits required to be fed from the normal source.

Sec. 517-19 — Critical care areas

(a) Patient bed location branch circuits. Each patient bed location must be supplied by at least one branch circuit supplied from the emergency system in addition to the one or more circuits fed from the normal system. At least one branch circuit from the emergency system must supply outlets only at that bed location. The emergency system receptacles are to be identified and must also indicate the panelboard and circuit number supplying them.

ARTICLE 517, PART C
ESSENTIAL ELECTRICAL SYSTEM
Sec. 517-25 — Scope

The essential electrical system must be capable of supplying the limited amount of lighting and power that is essential for life safety and orderly cessation of procedures during the time normal electrical service is interrupted for any reason.

Covered by this rule are clinics, medical and dental offices, outpatient facilities, nursing homes, limited care facilities, hospitals, and other health care facilities serving patients.

An FPN refers to NFPA 99, *Health Care Facilities* for more information on the need for an essential electrical system. The requirements there are discussed later in this chapter.

Sec. 517-30 — Essential electrical systems for hospitals

(a) Applicability. The requirements of Secs. 517-30 through -35 apply to hospitals where an essential electrical system is required.

FPNs again refer to NFPA 99 and, in addition, also to NFPA 20, *Installation of Centrifugal Fire Pumps.* The information contained in NFPA 20 will be summarized in Chapter 9 of this book.

(b) General. Essential electrical systems for hospitals must be comprised of two separate systems (**Fig. 8.2**) capable of supplying a limited amount of lighting and power that is essential for life safety and effective hospital operation during the time the normal electrical service is interrupted for any reason. These two systems are the:

emergency system, which is limited to supplying "life-safety branch circuits" essential to life safety, and the "critical branch" supplying power needed for critical patient care; and

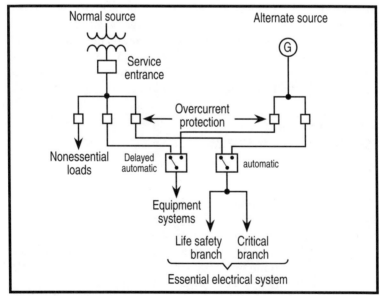

■ **Fig. 8.2.** Typical arrangement of transfer switches of a small hospital electrical system. Adapted from NEC Diagram 517-30(A).

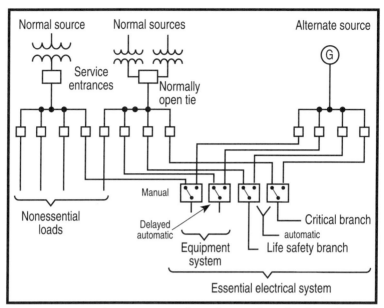

- **Fig. 8.3.** Typical arrangement of transfer switches of a large hospital electrical system. Adapted from NEC Diagram 517-30(B).

equipment system that supply major electric equipment necessary for patient care and basic hospital operation.

This is a somewhat shortened version of the rule given in NFPA 99. Information about the application of transfer switches in essential electrical systems, however, is more detailed in the NEC than in NFPA 99.

The number of transfer switches used are to be based upon reliability, design, and load considerations. Each branch of the essential electrical system must be served by one or more transfer switches, as shown in **Fig. 8.3**.

One transfer switch (**Fig. 8.4** on the next page) is permitted to serve one or more branches or systems in a facility with a maximum demand on the essential electrical system of 150kVA.

Loads other than essential electrical systems are permitted to be carried by the standby generator. These nonessential loads must be

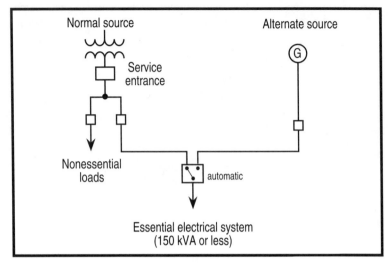

■ **Fig. 8.4.** Typical use of a single transfer switch in a small hospital, nursing home, and residential custodial care electrical system. Adapted from NEC Diagrams 517-30(C) and 517-41(C).

fed from their own transfer switch (**Fig. 8.5**). If the nonessential loads will overload the generating equipment when the essential loads are transferred to the generator upon loss of normal power, this switch is prohibited from transferring the nonessential load. Alternately, excess nonessential load must be shed when the generating equipment overloads.

An FPN refers to NFPA 99, Sec. 3-5.1.2.2(a) for transfer switch operation Type I; to Sec. 3-4.2.1.4 for automatic transfer switch features; and to Sec. 3-4.2.1.6 for nonautomatic transfer switch features. These are covered later this chapter.

(c) Wiring requirements. There are specific rules that apply to the separation and protection of essential electrical systems.

Emergency system. The *life-safety branch* and the *critical branch* of the emergency system must be kept entirely independent of all other wiring and equipment and must not enter the same raceways, boxes, or cabinets with each other or other wiring, except in:

- transfer switch enclosures;

- exit or emergency lighting fixtures supplied from two sources;
- a common junction box attached to exit or emergency lighting fixtures supplied from two sources; or
- wiring of two or more emergency circuits supplied from the same branch.

The wiring of the emergency system of a hospital must be mechanically protected by installation in nonflexible metal raceways, or wired with Type MI cable. There are, however, exceptions to this rule.
- Exception No. 1. Flexible power cords of appliances, or other utilization equipment, connected to the emergency system is not required to be enclosed in raceways.
- Exception No. 2. Secondary circuits of transformer-powered communication or signaling systems are not required to be enclosed

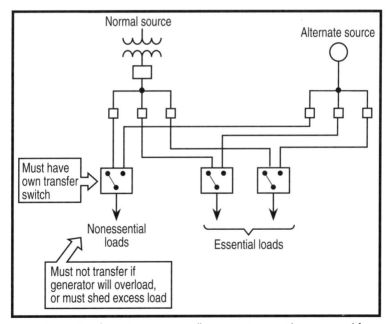

■ **Fig. 8.5.** An alternate source standby generator must be prevented from overloading.

in raceways unless otherwise specified in Chapter 7 or 8 of the NEC.
• Exception No. 3. Schedule 80 rigid nonmetallic conduit is permitted except for branch circuits serving patient care areas.
• Exception No. 4. Where encased in not less than 2 inches (50.8 mm) of concrete, Schedule 40 PVC is permitted except for branch circuits serving patient care areas.
• Exception No. 5. Type MI cable is permitted. Where installed as branch-circuit conductors serving patient care areas, the cable sheath must be identified as an acceptable grounding return path.
• Exception No. 6. Flexible metal raceways and cable assemblies are permitted in prefabricated medical headwalls or where necessary for flexible connection to equipment.

Equipment system. The wiring of the equipment system is permitted to occupy the same raceways, boxes, or cabinets of other circuits that are not part of the emergency system.

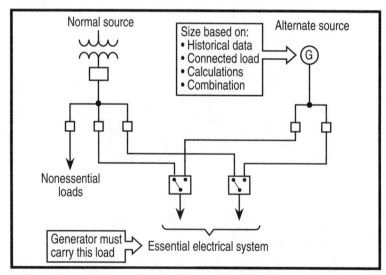

■ **Fig. 8.6.** It is essential that the alternate source generators be sized properly, neither too small to carry the load, nor too large when they are engine driven.

(d) Capacity of systems. The essential electrical system must have adequate capacity to meet the demand for the operation of all functions and equipment to be served by each system and branch. Feeders must be sized in accordance with **Articles 215** and **220**. The generator set(s) must have sufficient capacity and proper rating to meet the demand produced by the load of the essential electrical system at any one time (**Fig. 8.6**). Demand calculations for sizing the generator set(s) are to be based on:
- prudent demand factors and historical data; or
- connected load; or
- feeder calculation procedures described in **Article 220**; or
- any combination of these factors.

Correctly sizing an emergency standby generator is essential. Significantly oversizing is detrimental to engine performance and also not in accordance with the required testing protocols in NFPA 99.

Sec. 517-31 — Emergency system

Those functions of patient care depending on lighting or appliances that are connected to the emergency system are to be divided into two mandatory branches: the *life safety branch*; and the *critical branch*. The requirements of each are described more at length in Secs. 517-32 and -33.

The branches of the emergency system must be installed and connected to the alternate power source so that all functions specified for the emergency system will be automatically restored to operation within *10 seconds* after interruption of the normal source.

Sec. 517-32 — Life safety branch

No function other than those listed below are allowed to be connected to the life safety branch.

The life safety branch of the emergency system must supply power for the following lighting, receptacles, and equipment.
- Illumination of means of egress, such as lighting required for corridors, passageways, stairways and landings at exit doors, and all necessary ways of approach to exits. An FPN refers to NFPA

101, Secs. 5-8 and 5-9 for more information on this subject. It is discussed later in this chapter.
 • Exit signs and exit directional signs. An FPN refers to NFPA 101, Sec. 5-10 for more details. These are discussed later in this chapter.
 • Alarms and alerting systems including fire alarms (NFPA 101, Secs 7-6 and 12-3.4), and alarms required for systems used for the piping of nonflammable medical gases (NFPA 99, Section 12-3.4.1). These are discussed later in this chapter.
 • Communication systems in hospitals, where used for issuing instructions during emergency conditions.
 • Generator set location selected receptacles, plus the task-illumination battery charger for emergency battery-powered lighting unit(s) in the area.
 • Elevator cab lighting, control, communication, and signal systems.

Note that switching of patient corridor lighting in hospitals is permitted for the purpose of transferring from general illumination circuits to night illumination circuits. However, the switching arrangement must have only two positions (DAY or NIGHT) and have no OFF position that would allow both circuits to be extinguished at the same time.

Sec. 517-33 — Critical branch

(a) Task illumination and selected receptacles. The critical branch of the emergency system must supply power for task illumination, fixed equipment, selected receptacles, and special power circuits serving areas and functions related to patient care such as:
 • task illumination, selected receptacles, and fixed equipment in critical care areas that utilize anesthetizing gases;
 • isolated power systems in special environments;
 • task illumination, all receptacles, and fixed equipment in infant nurseries, medication preparation areas, pharmacy dispensing areas, selected acute nursing areas, psychiatric bed areas (omit receptacles), ward treatment rooms; and nurses' stations (unless adequately lighted by corridor luminaires;

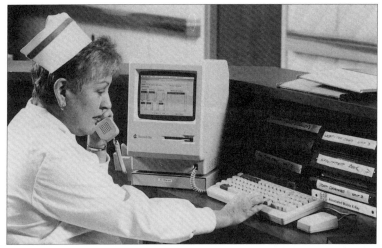

Nurse call and nurse patient communication systems must be connected to the critical branch of the emergency system.

• other specialized patient care task illumination and receptacles, where needed;
• nurse call systems;
• blood, bone, and tissue banks;
• telephone equipment room and closets;
• task illumination, selected receptacles, and selected power circuits for general care beds (at least one duplex receptacle per patient bedroom), angiographic labs, cardiac catheterization labs, coronary care units hemodialysis rooms or areas, selected emergency room treatment areas, human physiology labs, intensive care units, and selected postoperative recovery rooms; and
• additional task illumination, receptacles, and selected power circuits needed for effective hospital operation. Single-phase fractional horsepower exhaust fan motors that are interlocked with 3-phase motors on the equipment are permitted to be connected to the critical branch.

(b) Subdivision of the critical branch. It is permitted to subdivide the critical branch into two or more branches.

An FPN points out that is important to analyze the consequences of supplying an area with only critical care branch power when failure occurs between the area and the transfer switch. Dividing the normal and critical power, or feeding critical power from separate transfer switches, may be appropriate for increasing reliability.

Sec. 517-34 — Equipment system connection to alternate power source

The equipment system must be installed and connected to the alternate source, such that the equipment is automatically restored to operation at appropriate time-lag intervals following the energizing of the emergency system. The following sequence (**Fig. 8.7**) is to be maintained.

(a) Equipment for delayed automatic connection. The following equipment must be arranged for delayed automatic connection to the alternate power source.

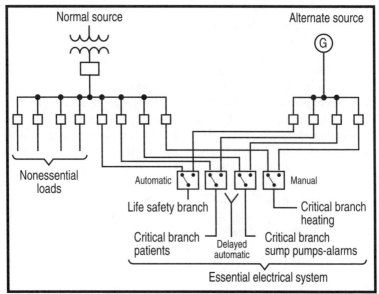

■ **Fig. 8.7.** Requirements for transfer switches in a hospital.

- Central suction systems serving medical and surgical functions, including controls. Such suction systems are permitted on the critical branch.
- Sump pumps and other equipment required to operate for the safety of major apparatus, including associated control systems and alarms.
- Compressed air systems serving medical and surgical functions including controls.
- Smoke control and stair pressurization systems.
- Kitchen hood supply and/or exhaust systems if required to operated during a fire in or under the hood.

An exception says that the above equipment may be arranged for sequential delayed automatic connection to the alternate power source to prevent overloading the generator where engineering studies indicate it is necessary.

(b) Equipment for delayed automatic or manual connection. The arrangement must also provide for the subsequent connection of the following equipment to the alternate power source. This equipment can be either delayed automatic or manually connected.
- Equipment to provide heating for operating, delivery, labor, recovery, intensive care, coronary care, nurseries, infection/isolation rooms, emergency treatment spaces, and general patient rooms.

An exception to this requirement says that heating of general patient and infection/isolation rooms are not required if the outside design temperature is higher than +20°F (-6.7°C); or where a selected heated room(s) is provided for the needs of all confined patients; or if the facility is served by a dual source of normal power, which give the service greater than normal reliability.

An FPN to Sec. 517-34(b) says that the design temperature is based on the $97^1/_2\%$ design value as shown in Chapter 24 of ASHRAE *Handbook of Fundamentals*.
- Elevator(s) selected to provide service to patient, surgical, obstetric, and ground floors during interruption of normal power. In instances where interruption of normal power would result in other elevators stopping between floors, throw-over facilities must

be provided to allow the temporary operation of any elevator for the release of patients or other persons who may be trapped between floors.

• Supply, return, and exhaust ventilating systems for surgical and obstetric delivery suites, intensive care, coronary care, nurseries, infection/isolation rooms, emergency treatment spaces, and exhaust fans for laboratory fume hoods, nuclear medicine areas where radioactive material is used, ethylene oxide evacuation, and anesthesia evacuation.

• Hyperbaric and hypobaric facilities.
• Automatically operated doors.
• Minimal electrically heated autoclaving equipment is permitted to be arranged for either automatic or manual connection to the alternate source.
• Other selected equipment is permitted to be served by the equipment system.

Sec. 517-35 — Sources of power

(a) Two independent sources of power. Essential electrical systems must have a minimum of two independent sources of power: a normal source generally supplying the entire electrical system, and one or more alternate sources for use when the normal source is interrupted.

(b) Alternate source of power. The alternate source of power must be a generator(s) driven by some form of prime mover(s), and located on the premises.

An exception to this rule is made for cases where the normal source consists of generating units on the premises. In that event, the alternate source can be either another generating set, or an external utility service.

(c) Location of essential electrical system components. Careful consideration must be given to the location of the spaces housing the components of the essential electrical system to minimize interruptions caused by natural forces common to the area (storms, floods, earthquakes, or hazards created by adjoining structures or activities).

Consideration must also be given to the possible interruption of normal electrical services resulting from similar causes as well as possible disruption of normal electrical service due to internal wiring and equipment failures.

An FPN notes that facilities whose normal source of power is supplied by two or more separate central-station-fed services experience greater-than-normal electrical service reliability than those with only a single feed. Such a dual source of normal power consists of two or more electrical services fed from separate generator sets or a utility distribution network having multiple power input sources and arranged to provide mechanical and electrical separation so that a fault between the facility and the generating sources will not likely cause an interruption of more than one of the facility service feeders.

Sec. 517-40 — Essential electrical systems for nursing homes and limited care facilities

Nursing homes and limited care facilities that provide inpatient hospital care must comply with the requirements of Part C, Secs. 517-40 through -44 that follow.

An exception is made for freestanding buildings used as nursing homes and limited-care facilities, provided that:

- it maintains admitting and discharge policies that preclude the provision of care for any patient or resident who may need to be sustained by electrical life-support equipment; and
- offers no surgical treatment requiring general anesthesia; and
- provides an automatic battery-operated system or equipment that is effective for at least $1^1/_2$ hrs and meets the requirements of Sec. 700-12. The system must be capable of supplying lighting for exit lights, exit corridors, stairways, nursing stations, medical preparation areas, boiler rooms, and communication areas, and supply power to operate all alarm systems.

Note that nursing homes and limited care facilites that do provide inpatient hospital care must instead comply with the rules applicable to *essential systems for hospitals* (Part C, Secs. 517-30 through -35).

In addition, nursing homes and limited care facilities that are contiguous with a hospital are permitted to have their essential electrical systems supplied by that of the hospital.

An FPN refers to NFPA-99 for performance, maintenance, and testing requirements for essential electrical systems in nursing homes and limited care facilities. The reference here is to Sec. 3-4.2.3 which covers essential electrical distribution requirements for Type II systems. This is discussed later in this chapter.

Sec. 517-41 — Essential electrical systems (nursing homes, etc.)

(a) General. In many ways the requirements as similar to those for hospitals. Essential electrical systems for nursing homes and limited care facilities are to be made up of two separate branches capable of supplying a limited amount of lighting and power service that is considered essential for the protection of life safety and effective operation of the institution during the time normal electrical service is interrupted for any reason. These two separate branches are: the <u>life safety branch</u> and the <u>critical branch</u>. In these facilities, there is no requirement for equipment systems as in hospitals.

It should be noted that NFPA 99 calls the life safety branch the "emergency system," and the critical branch is called the "critical system."

(b) Transfer switches. The number of transfer switches used is to be based upon reliability, design, and load considerations. Each branch of the essential electrical system must be served by one or more transfer switches. Similar to hospitals, one transfer switch is permitted to serve one or more branches or systems in a facility with a maximum demand on the essential electrical system of 150kVA.

An FPN refers to NFPA 99, Secs. 3-5.1.2.2(b) for transfer switch operation Type II; 3-4.2.1.4 for automatic transfer switch features; and 3-4.2.1.6 for nonautomatic transfer device features.

(c) Capacity of system. The essential electrical system must have adequate capacity to meet the demand for the operation of all functions and equipment to be served by each branch at one time.

(d) Separation from other circuits. The life safety branch must be kept entirely independent of all other wiring and equipment and must not enter the same raceways, boxes, or cabinets with other wiring except in:
- transfer switches;

- exit or emergency lighting fixtures supplied from two sources; or
- a common junction box attached to exit or emergency lighting fixtures supplied from two sources.

The wiring of the critical branch is permitted to occupy the same raceways, boxes, or cabinets of other circuits that are not part of the life safety branch.

Sec. 517-42 — Automatic connection to life safety branch

The life safety branch must be so installed and connected to the alternate source of power that all functions specified must be automatically restored to operation within 10 seconds after the interruption of the normal source. An FPN notes that the life safety branch is called the emergency system in NFPA 99.

No function other than those listed below are to be connected to the life safety branch. The life safety branch must supply power for the following lighting, receptacles, and equipment.

- Illumination of the means of egress as necessary for corridors, passageways, stairways, landings, and exit doors and all ways of approach to exits. As with hospitals, an FPN refers to NFPA 101, Secs. 5-8 and 5-9 for more information on the subject.
- Exit signs and exit directional signs. NFPA 101, Sec. 5-10 provides more details.
- Alarms and alerting systems. Fire alarms per NFPA 101, Secs. 7-6 and 12-3.4; and those used for systems used for the piping of nonflammable medical gases (NFPA 99, Sec. 16-3.4.1).
- Communications systems where used for issuing instructions during emergency conditions.
- Dining and recreation area lighting sufficient to provide illumination to exit ways.
- Generator set location task illumination and selected receptacles.
- Elevator cab lighting, control, communication, and signal systems.

Note that switching of patient corridor lighting is permitted for the purpose of transferring from general illumination circuits to others (such as night lighting circuit), but the switching arrangement must have only two positions and have no OFF position that would allow both circuits to be extinguished at the same time.

Sec. 517-43 — Connection to critical branch

The critical branch must be so installed and connected to the alternate power source that the equipment is automatically restored to operation at appropriate time-lag intervals following the restoration of the life safety branch to operation.

- *(a) Delayed automatic connection.* The following equipment connected to the critical branch is to be arranged for delayed automatic connection to the alternate power source.
- Patient care area task illumination and selected receptacles in medication preparation areas, pharmacy dispensing areas, and nurses' stations (unless adequately lighted by corridor luminaires).
- Sump pumps and other equipment required to operate in order to assure the safety of major apparatus and associated control systems and alarms.
- Smoke control and stair pressurization systems.
- Kitchen hood supply and/or exhaust systems required to operate during a fire in or under the hood.

(b) Delayed automatic or manual connection. The circuit arrangement must also provide for the additional connection of other equipment by either delayed automatic or manual operation. The following equipment must be connected to the critical branch.

- Heating equipment to provide heating for patient rooms. An exception says that heating of general patient rooms is not required if the outside design temperature is higher than +20°F (-6.7°C), or where a selected room(s) is provided for the needs of all confined patients, then only such room(s) need to be heated, or the facility is served by a dual source of normal power that provides greater than normal electrical service reliability.
- Elevator service. In instances where disruption of power would result in elevators stopping between floors, throw-over facilities must be provided to allow the temporary operation of any elevator to release passengers.

- Additional items such as illumination, receptacles, and equipment are permitted to be connected only to the critical branch.

Sec. 517-44 — Sources of power

(a) Two independent sources of power. Essential electrical systems must have a minimum of two independent sources of power: a normal source generally supplying the entire electrical system, and one or more alternate sources for use when the normal source is interrupted.

Centralized inverters are often used to supply an alternate source of power to exit lights, illumination, and other loads that must be maintained in operation during an outage of the normal source.

(b) Alternate source of power. The alternate source of power must be a generator(s) driven by some form of prime movers(s), and located on the premises. There are exceptions to this rule.

Exception No. 1: Where the normal source consists of generating units on the premises, the alternate source can be either another generator set, or an external utility service.

Exception No. 2: Nursing homes or residential custodial care facilities meeting the requirements of Sec. 517-40(a), Exception are permitted to use a battery system of self-contained battery integral with the equipment.

(c) Location of essential electrical system components. Similar to hospitals, careful consideration should be given to the location of components of the essential electrical system to minimize interruptions caused by natural forces common to the area, and also to possible interruption of normal electrical services due to internal wiring and equipment failures. Facilities whose normal source of power is supplied by two or more separate central-station-fed services experience greater-than-normal service reliability.

Sec. 517-50 — Systems for other health care facilities

This section contains rules for the essential electrical systems for clinics, medical and dental offices, outpatient facilities, and other health care facilities not covered by Secs. 517-30, -40, and -45.

Connections. The essential electrical system must supply power for the following.

- Task illumination that is related to the safety of life and that is necessary for the safe cessation of procedures in progress.
- All anesthesia and resuscitative equipment used in areas where inhalation anesthetics are administered to patients, including alarm and alerting devices. An FPN refers to NFPA 99, Sections 13-3.4.1, 14-3.4.1, and 15-3.4.1 for more information on the subject.

Alternate source of power. The alternate source of power for the system must be specifically intended for this purpose and must be either a generator, battery system, or self-contained battery integral with the equipment. In the case where electrical life-support equipment is required, an exception says that the essental electrical system must comply with Secs. 517-30 through -35.

The alternate source of power is to be separate and independent of the normal source and have a capacity to sustain its connected loads for a minimum of 1½ hours after the loss or the normal source.

The alternate source of power must be so arranged that, in the event of a failure of the normal power source, the alternate source of power is to be automatically connected to the load within 10 seconds.

An FPN refers to NFPA 99, Sections 3-5.1.2.2(c) for information on transfer switch operation for Type III with generator sets; and 3-5.1.2.2(d) for information on transfer switch operation for Type III with battery systems.

NFPA Standards

More than in any other articles of the NEC, those that deal with emergency, standby, and other types of onsite power make references to NFPA and other standards. Many of the details on these subjects contained in these document are crucial to the full understanding of the intent of some of the NEC rules. For that reason, some of their most important provisions are included here. Before applying any of them to a specific electrical installation, however, it is essential that the pertinent parts of the latest issue of these documents be read in their entirety.

NFPA 99 — Standard for Health Care Facilities (1993 edition). Chapter 3 of this document contains the requirements for electrical systems installed in health care facilities. Specifically, **Sec. 3-4.2** covers "essential systems." Three types of these systems are identified.

Type I essential electrical systems apply to those required for hospitals. The definition is exactly the same as that given in NEC Sec. 517- 30(b).

References are made in Sec. 517-30(b) to **Sec. 3-5.1.2.2** for Type I transfer swithch operation. There it says that the essential electrical sysem must be served by the normal power source except when the normal power source is interrupted or drops below a predetermined voltage level. Settings of the sensors are to be determined by careful study or the voltage requirements of the load.

Failure of the normal source must automatically start the alternate source generator after a short delay. When the alternate source has attained a voltage and frequency that satisfies minimum operating requirements of the essential electrical system, the load is to be connected automatically to the alternate power source.

Upon connection of the alternate power source, the loads comprising the emergency system must be automatically reenergized. The load comprising the equipment system is to be connected either automatically after a time delay or nonautomatically and in such a sequential manner as not to overload the generator.

When the normal power source is restored, and after a time delay, the automatic transfer switches must disconnect the alternate source of power and connect the loads to the normal power source. The alternate power source generator set is to continue to run unloaded for a preset time delay.

If the emergency power source fails and the normal power source has been restored, retransfer to the normal source of power must be immediate, bypassing the retransfer delay timer.

If the emergency power source fails during a test, provisions must be made to immediately retransfer to the normal source.

Nonautomatic transfer switching devices must be restored to the normal power source as soon as possible after the return of the normal source at the discretion of the operator.

Sec. 517-32(b) also refers to **Sec. 3-4.2.1.4** for automatic transfer switch features. Generally, they must be electrically operated and mechanically held, and transfer and retransfer automatically. An exception is made where it is desirable to program the transfer switch for a manually initiated retransfer to the normal source, or for an automatic intentional off delay, or for an in-phase monitor relay or similar automatic delay method, to provide for a planned momentary interruption of the load. If used, a bypass feature must permit automatic retransfer in the event that the alternate source fails and the normal source is available.

The remaining items listed in this section of NFPA 99 details items the manufacturer must include in the design of the automatic transfer switch.

In addition, Sec. 517-32(b) refers to **Sec. 3-4.2.1.6** for features of nonautomatic transfer devices. They must be mechanically held and the operation must be by direct manual or electrical remote manual control. Electrically operated switches must derive their control power from the source to which the load is being transferred. A means for safe manual operation is to be provided. The remaining items listed detail items the manufacturer must include in the design of the switch.

Sec. 517-32(c) refers to **Sec. 12-3.4.1** of the standard for information on alarms for systems carrying nonflammable medical gases. This section in turn refers back to Secs. 4-3 through 4-6. Sec. 4-4.1.1 deals with gas warning systems and requires a master alarm system to monitor the operation and condition of the source of supply, the reserve, and pressure in the main line of the medical gas system. This master alarm panel is to be independent of a computer or building management system. An area alarm system is also required in anesthetizing and other critical care location to monitor the pressure in the local supply line. Other requirements are scattered throughout these sections.

Sec. 517-40 refers to the standard as it applies to essential electrical systems in nursing homes and limited care facilities. An essential definition here

Type II essential electrical systems are those used in nursing homes and limited care facilities. The requirements here basically match those described in Sec. 517-40.

Sec. 517-41(b) refers to **Sec. 3-5.1.2.2(b)** for information on transfer switch operation in a Type II application. There it says that the essential electrical electrical system must be served by the normal power source except when the normal power source is interrupted or drops below a predetermined voltage level. Settings of the sensors is to be determined by careful study of the voltage requirements of the load.

Failure of the normal source must automatically start the alternate source generator after a short delay. When the alternate power source has attained a voltage and frequency that satisfies minimum operating requirements of the essential electrical system, the load is to be connected automatically to the alternate power source.

Upon connection of the alternate power source, the loads comprising the emergency system (called the life safety branch in the NEC) are to be connected either automatically after a time delay or nonautomatically and in such a sequential manner as not to overload the generator.

When the normal power source is restored, and after time delay, the automatic transfer switches must disconnect the alternate source of power and connect the loads to the normal power source. The alternate power source generator set is to continue to run unloaded for a preset time delay.

If the emergency power source should fail and the normal power source has been restored, retransfer to the normal source of power is to be immediate, bypassing the retransfer delay timer.

If emergency power source fails during a test, provisions must be made to immediately retransfer to the normal source.

Nonautomatic transfer switching devices must be restored to the normal power source as soon as possible after the return of the normal source or at the discretion of the operator.

Type III essential electrical systems are those found in clinics, medical and dental offices, and health care facilites other than hospitals, nursing homes, and limited care facilities. The requirements here are similar to those covered in NEC Sec. 517-50.

Sec. 517-50(c) refers to Sec. 3-5.1.2.2(d) for more information on transfer switch operation for Type III systems with battery systems.

In this case, failure of the normal source must automatically transfer the load to the battery system. Retransfer to the normal source must be automatic upon restoration of the normal source.

NFPA 101 — Life Safety Code

Secs. 5-8 and 5-9 contain information on the illumination of the means of egress, while **Sec. 5-10** specifies the requirements for exit and exit directional signs. **Secs. 7.6** and **12-3.4** deal with alarm and alerting systems.

All these sections are referred to as part of the rules given in the NEC for the auxiliary systems for health-care facilities. The provisions of NFPA 101 are very important because details on these items are not included in the NEC. Because it is essential that the requirements of the latest issue of the Life Safety Code be adhered to, they will not be discussed here. The 1996 NEC refers to the 1994 issue of NFPA 101. Be sure to seek out the most recent issue of the standard when designing such systems.

Firepumps motors and other equipment that make up a fire protection system must comply with the requirements of the NEC and NFPA standards.

ARTICLE 695 — FIRE PUMPS

9.

Safety is increased and fire-insurance rates are reduced when a sprinkler system is included in buildings and manufacturing facilities. Fire pumps, either electric-motor driven or engine driven, are an integral part of the fire-protection system. They are used where the water requirements, which can be in the thousands of gallons per minute range at high pressures, cannot be met by public supplies or gravity tanks.

Fire pumps are covered in this book because they are linked in many ways to auxiliary power generating systems. Motor-driven units must be supplied with an alternate power source if the utility source is not reliable. Engine-driven units are, in effect, dedicated mechanical sources. NFPA 20 that covers fire pumps is a referred-to document in Article 517 (Chapter 8). Only the most important sections of Article 695 and NFPA 20 will be discussed here.

Sec. 695-1 — Scope

This article covers the installation of electrical power sources and interconnecting circuits, and switching and control equipment dedicated to fire pump drivers. The drivers can be either motors or diesel engines.

What is not covered by this article is the performance, maintenance, and acceptance testing of the firepump system, the internal wiring of the components of the system, and the pressure maintenance (jockey or make-up) pumps.

More information on fire pumps is to be found in the latest issue of NFPA 20, *Standard for the Installation of Centirfugal Fire Pumps*. The important requirements in that document is summarized at the end of this chapter.

Sec. 695-3 — Power source to motor driven fire pumps

Power to a fire pump must be supplied to an electric motor driven fire pump by either an electric utility service or onsite generation.

When supplied by onsite generation, the generation facility must be located and protected to minimize the possibility of being damaged by fire.

The supply conductors must connect the power source (either a service or onsite generation) directly to a listed firepump controller, as shown in **Fig. 9.1(A)**. There are, however, exceptions to this rule.

Exception No. 1 permits a disconnecting means and overcurrent protective device to be installed between the power supply and controller, but these items must comply with the following.

- The overcurrent device must be selected or set to carry indefinitely the sum of the locked-rotor current of the firepump motor(s) and the jockey-pump motor(s), and the full-load current of associated firepump accessory equipment connected to firepump power supply.
- The disconnecting means must be marked as being suitable for use as service equipment and must be lockable in the ON position. Note that this does not mandate the locking of the disconnect; a later provision allows other alternatives.
- A placard must be installed on the exterior of the disconnecting means stating "FIRE PUMP DISCONNECTING MEANS" in letters at least 1-in. (25.4 mm) high.
- A placard must be placed adjacent to the firepump controller stating the location of this disconnecting means (and the location of the key if the disconnecting means is locked in the ON position).
- The disconnecting means must be supervised in the closed position by either: a central station, proprietary, or remote station signal device; a local signaling service that will cause the sounding of an audible signal at a constantly attended point; locking the disconnecting means closed; or sealing the disconnecting means and approved weekly *recorded* inspections (only if the disconnecting means is within a fenced enclosure or in a building under the control of the owner).

Exception No. 2 permits a transformer to be installed ahead of the firepump controller in cases where the supply voltage is different from the utilization voltage, as shown in **Fig. 9.1(B)**. The transformer must meet the requirements of Sec. 695-5, and a disconnecting means and overcurrent protective devices for protecting the transformer are also permitted providing they meet the requirements of Ex. 1 discussed previously.

Sec. 695-4 — Multiple power sources for firepump motors

Where reliable power cannot be obtained from a electric utility or onsite source, the power to the fire pump is to be supplied by either:
- two or more electric utility or onsite sources in combination

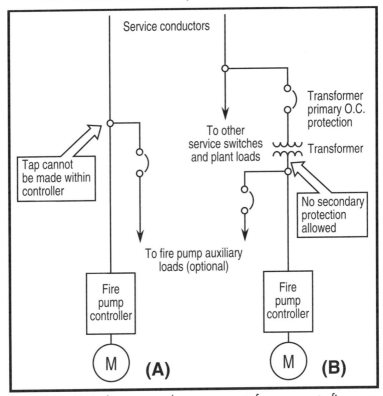

■ **Fig. 9.1.** Typical power supply arrangements from source to fire pump motor. **(A)** is for instances where the utility or onsite generation is at the voltage required by the firepump motor. This is "Arrangement A" in NFPA 20. **(B)** is for instances where the service voltage must be reduced by a transformer to the utilization voltage of the firepump motor. This is "Arrangement B" in NFPA 20.

(could be a double-ended substation); or
- a utility or onsite source in combination with an onsite generator.

The multiple power source must be arranged so that a fire at one source will not cause an interruption at the other source. In addition, it must be approved by the authority having jurisdiction.

The supply conductors from the multiple power source must directly connect to either a listed combination firepump controller and power transfer switch, or to a disconnecting means and overcurrent protective device meeting the requirements of Sec. 695-3(c), Ex. 1. Note, however, that as shown in **Fig. 9.2**, the transfer switch supplied by the firepump manufacturer can be either integral within the con-

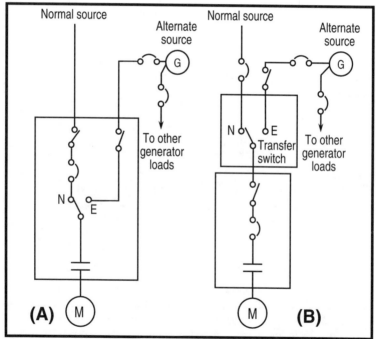

■ **Fig. 9.2.** Transfer switches for firepump service are supplied with the firepump controller, either within the controller **(A)** (referred to in NFPA 20 as "Arrangement A"), or as a separate unit **(B)**, which is "Arrangement B."

troller (**A**), or as a separate unit (**B**). Both are acceptable.

Where one of the alternate power sources is an onsite generator, the disconnecting means and overcurrent protective device for these supply conductors must be selected or set to allow instantaneous pickup and running of the full pump-room load.

Sec. 695-5 — Transformers

Sizing. Where a transformer is dedicated to supplying a motor-driven fire pump installation, it must be rated at a minimum of 125% of the sum of:
- the rated full load of the firepump motor(s); plus
- the rated full load of the pressure-maintenance pumps(s) when connected to this power supply; and
- the full load of any associated fire pump accessory equipment connected to this power supply.

Overcurrent protection for the transformer must comply with the following.

Primary overcurrent protection is permitted to be selected or set as high as 600% of the rated full-load current of the transformer. This selection or setting must be sufficient to carry indefinitely the equivalent of the transformer secondary current sum of:
- the locked-rotor currents of the firepump motor(s); plus
- the locked-rotor currents of the pressure-maintenance pump motor(s) connected to this supply; and
- the full-load current(s) of any associated firepump accessory equipment when connected to this power supply.

Secondary overcurrent protection is not permitted.

Sec. 695-7 — Equipment location

Motor-driven fire pump. The controller and transfer switch must be located as close as practicable to the motors they control and must be within sight of the motors.

Engine-driven fire pump. The controller must be located as close as practicable to the engine they control and must be within sight of the engine.

Storage batteries for a diesel engine drive must be rack supported above the floor, secured against displacement, and located where they will not be subjected to excessive temperature, vibration, mechanical injury, or flooding with water.

Energized equipment parts must be located at least 12 in. (305 mm) above the floor level.

Protection. Controllers and transfer switches must be located or protected so that they will not be damaged by water escaping from pumps or pump connections.

Control equipment must be mounted in a substantial manner on noncombustible supporting structures.

Sec. 695-8 — Power wiring

Supply conductors must be physically routed outside buildings and be installed as service-entrance conductors in accordance with **Article 230**, as shown in **Fig. 9.3(A)**. Where supply conductors cannot be physically routed outside buildings, they are permitted to be routed through buildings where installed under, or enclosed within, not less than 2 in. (50.8 mm) of concrete (**B**).

Exception No. 1. Firepump supply conductors on the *load side* of the disconnecting means permitted by Sec. 695-3(c), Ex. 1, are permitted to be routed through the building using listed electrical circuit protective systems with a minimum of 1-hr fire resistance. The installation must comply with the restrictions provided for in the listing of such systems. Two listed items that meet these requirements are a 2-in. or larger rigid steel conduit wrapped with a fire-retardant wrapping, and Type MI cable.

Exception No. 2. When the supply conductors are run within, the electrical switchgear room in which they originate; or within the firepump room.

Wiring methods. Wiring from the controllers to the pump motors must be in:
- rigid metal conduit;
- intermediate metal conduit;
- liquidtight flexible metal conduit; or
- Type MI cable.

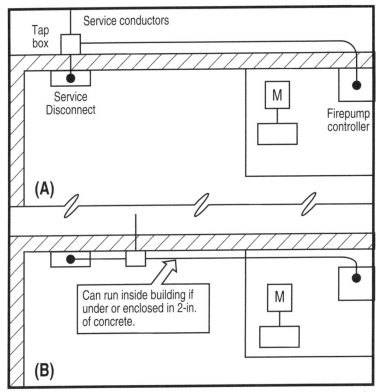

■ **Fig. 9.3.** The preferred method of routing service conductors to the firepump controller is to run the raceway outside the building or structure **(A)**. When that is not possible, it can be run inside **(B)**, but it must either be under or enclosed within not less than 2 inches of concrete. The run within the firepump room does not have to be protected.

This requirement is illustrated in **Fig. 9.4** on the next page. As will be described later, the same requirement hold for control wiring that is part of the firepump system.

Conductors must only be protected against short circuit. In addition, they are to be protected as required or permitted by the following:

- **Sec. 230-90(a), Exception No. 4** requires the overcurrent protection to be selected or set to carry locked-rotor current of the

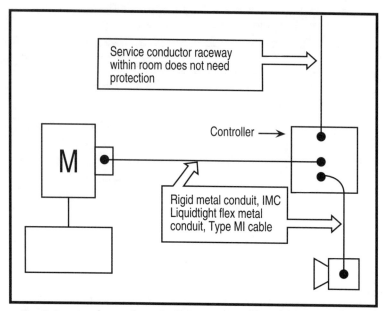

- **Fig. 9.4.** Conductors from the firepump controller to the motor, and from the controller to associated controls are critical. They, therefore, must be protected by running in either rigid metal conduit, IMC, liquidtight flexible metal conduit, or Type MI cable.

---

motor(s) indefinitely when the service to the firepump room is judged to be outside the building.

Voltage at the line terminals of the controller must not drop more than 15% below normal (controller-rated voltage) under motor starting conditions. The voltage at the motor terminals must not drop more than 5% below the voltage rating of the motor when the motor is operating at 115% of the full-load current rating of the motor.

This voltage requirement does not apply to mechanical starting during an emergency run of the equipment.

Installation requirements. All wiring from engine controllers and batteries must be installed in accordance with the controller and engine manufacturer's instructions. Such wiring must be protected against mechanical injury.

STANDARD NFPA 20

NFPA 20, *Standard for the Installation of Centrifugal Fire Pumps* is an essential document that must be used in concert with the requirements of NEC Article 695. It contains additional electrical requirements that must be met in order for a firepump system to be acceptable to the authority having jurisdiction.

Here is a summary of some of the most important electrical items not covered by the rules given in Article 695. It is presented here only as an aid to understanding the magnitude of the information contained in NFPA 20, and is not to be used as a substitute for the *latest revision* of that document. These items are from the 1993 issue of the standard.

General

Equipment protection. The location of the firepump must be kept above 40°F (5°C), and adequate ventilation is required. Artificial light must be provided at the pump location, and emergency lights must also be provided by fixed or portable battery-operated units connected to normal lighting circuits.

Complete testing of the control systems of firepump installations is required as part of a field acceptance test.

Field acceptance test. When the installation is completed, an operating test is to be made in the presence of the customer, the pump manufacturer (who also supplies the motor/engine and controller), and the authority having jurisdiction.

Operation. Before starting the unit for the first time after installation, all field-installed electrical connections must be checked. The motor must be momentarily operated to ensure it rotates in the proper direction.

Power supply (for high-rise buildings). Where electric motors are used and the height of the structure is beyond the pumping capability of the fire department apparatus, a reliable emergency source of power must be provided. The emergency source is permitted to be either a standby engine-driven fire pump or other emergency power source for services essential to the safety and welfare of high-rise building occupants.

Type MI cable is an allowed wiring method. Such a system has a 2-hour rating, and easily meets the 1-hr fire resistance rating requirement. However, it must be installed to more stringent requirements than those found in NEC Article 330. For example, it must be supported (using specified methods) every 3 ft and must be terminated at least 1 ft within the boundary of a protected area. Where power to fire pumps is concerned, such a system is much easier to install than a concrete-encased arrangement.

Motors. All motors must: be specifically listed for firepump service; be rated for continuous duty; applied only at voltages of ±10% of motor nameplate voltage; and comply with NEMA Standard MG-1 and NEMA Design B standard markings.

Onsite power generator systems. Where used to supply power to firepump motors while supplying other loads, automatic shedding of the loads not required for fire protection is required immediately prior to starting the fire pump(s). Automatic sequencing of the fire pumps is also allowed.

Transfer of power must take place within the pump room. Protective devices in the onsite power source circuits at the generator must allow instantaneous pickup of the full pump room load.

Starting and control. The fire pumps are to start via a pressure-actuated switch that is responsive to the water pressure in the fire protection system. A means must be provided to allow relieving the pressure to the switch to allow testing of the controller and fire pump. Each controller (including the jockey pump controller) must have their individual pressure-sensing line.

Where more elaborate fire-control systems are installed, fire pumps can be started by the fire-protection equipment (via the opening of normally closed contacts) before the water pressure switches operate. Remote manual fire station switches are also permitted to start the fire pumps, but not to stop them.

The controller for each unit of multiple pump units must incorporate a sequential timing device to prevent more than one pump from starting simultaneously. The units must start at 5 to 10 second intervals, but failure to start must not prevent the next one in the sequence from starting. Interlocking must assure a pump providing suction pressure to another starts before the pump it supplies.

The controller is to have a manual handle or lever that mechanically closes the contactor independently of the electrical control circuit. The manual system is to be mechanically latched so that opening the contactor is to remain at the option of the operator. The contactor is to return to the OFF position if the operator releases the handle before it reaches the fully ON position.

Shutdown is to be accomplished as follows.

- When the fire pump has been manually started, the operation of a pushbutton on the outside of the controller enclosure is to return the controller to automatic operation.
- Where the mode of operation is automatic shutdown after automatic start, when the cause of the start has returned to normal, controls must require at least 10 minutes additional running time before pump shutdown.

Diesel-engine drive

Diesel engines have proved to be the most dependable internal combustion engine for driving fire pumps. Standard, spark-ignition internal combustion engines are not to be used for this purpose. This restriction does not apply to gas turbine engines.

Engines. Engines must be specifically listed for firepump service by a qualified testing laboratory.

Engines are to be supplied with an overspeed device set to shut down the engine at approximately 20% above rated engine speed. A means must be provided for the overspeed trouble signal to be interlocked so that the engine controller cannot be reset until the overspeed device is manually reset to the normal operating position. A rpm indicating tachometer must be provided. It must be able to record total time of engine operation, or a separate hour meter must be provided for this purpose.

All wiring of engine-mounted devices is to terminate in an engine junction box to terminals numbered to correspond with numbered terminals in the remote automatic controller. Interconnections between the junction box and controller are to be made using stranded wire sized on a continuous-duty basis.

The main battery contactors supplying current to the starting motor must be capable of manual mechanical operation to energize the starting motor in the event of control-circuit failure. Battery cables are to be provided per the engine manufacturer's recommendation. A speed-sensitive switch is to signal engine running and crank termination. Power for this signal is to be taken from a source other than the engine generator or alternator.

Engines must be equipped with a reliable starting device. Where electric starting is used, the device must be battery operated. If hydraulic starting is used, electrical means must automatically provide and maintain the stored hydraulic pressure within predetermined pressure limits.

All controls for engine shutdown in the event of low engine lubrication level, overspeed, and high jacket water temperature must be from a 12 or 24V DC source compatible with the controls on the engine. An interlock must be provided to prevent a hydraulic starting system for recranking until the interlock has been manually reset for automatic starting.

Each engine must be provided with two storage battery units. Lead-acid batteries are to be furnished in a dry-charge condition, with the electrolyte liquid in a separate container. This is to be added and a conditioning charge given at the time the engine is put in service. At 40°F (4.5°C), each battery unit must have sufficient capacity to maintain recommended cranking speed through a 6-min cycle (15-sec of cranking and 15-sec rest, in 12 consecutive tries).

Two means for recharging storage batteries must be provided. One means is via the engine generator or alternator, and the other via an automatically controlled charger taking power from an AC power source. An alternate charging source is permitted if the AC power source is not available or reliable.

The chargers must be specifically listed for firepump service, and the rectifier must be a semiconductor type. The charger for lead-acid batteries must be of the type that automatically reduces the charging rate to less than 500mA when the battery reaches a full charge condition. In addition, at rated voltage, the charger must be capable of delivering energy into a fully discharged battery in a manner that will not damage the battery. It must restore 100% of the battery's amp-hr rating within 24 hrs, and must automatically charge at the maximum rate whenever required by the state of charge of the battery. Where connected through a control panel, the charger must indicate loss of current output on the load side of the DC overcurrent protective device.

Storage batteries must be kept charged at all times. They must be tested frequently to determine the condition of the battery cells and the amount of charge in the battery. Only distilled water is to be used in battery cells, and the plates must be kept submerged at all times. Periodic inspection of the batteries and charger must determine that the charger is operating correctly, the water level in the battery is correct, and the battery is holding its proper charge.

Engine-drive controllers

In most respects, the requirements for the controller for a diesel-engine driven firepump controller are similar to those of one that is motor driven. Here are some items that are special requirements of diesel-engine driven fire pumps.

General. All controllers must be specifically listed for diesel-engine driven firepump service. They must be completely assembled, wired, and tested by the manufacturer before shipment from the factory.

Controllers must be marked "Diesel Engine Pump Controller" and show clearly the name of the manufacturer, identifying designations, and complete electrical rating. Where multiple pumps are provided serving different areas of a facility, a conspicuous sign must be attached on the outside of each controller indicating the portion of the system served by the fire pump.

It is the responsibility of the pump manufacturer to arrange for service and adjustment of the controller during the installation, testing, and warranty periods.

Components. Visible indicator alarms must be provided to indicate that the controller is in the "automatic" position. If the indicator is a pilot lamp, it must be accessible for replacement. Separate visible indicators and a common audible alarm capable of being heard while the engine is running and operable in all positions of the main switch except OFF must be provided to indicate trouble caused by the following conditions:

- critically low oil pressure in the lubrication system;
- high engine jacket coolant temperature;
- overspeed shutdown;
- battery failure;
- battery charger failure; and
- low air or hydraulic failure (where applicable).

No audible alarm silencing switch, other than the controller main switch is permitted for these alarms.

In addition, the following alarms are optional.

- low pumproom temperature;
- relief-valve discharge;
- flow meter left ON, bypassing the pump;
- water level in suction supply below normal;
- water level in suction supply near depletion;
- diesel engine supply below normal; and
- steam pressure below normal.

These alarms may be located at the controller or may be independent. When these alarms are used, a silence switch is to be incorporated, but the circuitry must cause the alarm to sound if the switch is in the SILENCE position while the supervised conditions are normal.

Centralized alarms for fire pumps are increasing in popularity as fire suppression systems become more important in hospitals and other facilities.

When the pump room is not constantly attended, audible of visible alarms powered by a source other then the engine starting batteries and not exceeding 125V must be provided at a point of constant attendance. These alarms must indicate:
- engine running;
- controller main switch in OFF or MANUAL position; and
- trouble on the controller or engine.

These alarms are to be derived from contacts in the controller. The controller must also be equipped with a pressure recording device that will operate for at least 24-hrs after the loss of its primary power source. This could involve a mechanically wound spring or a reliable electrical source.

Starting and control. To ensure dependable operation, the controller is to automatically start and run the engine for at least 30 minutes once a week. A solenoid valve drain on the pressure control line is to be the initiating means. The performance of this test is to be recorded as a pressure drop indication on a pressure recorder.

Starting is to be accomplished as follows:

- Power to manually and automatically start the engine is to be derived from the two storage battery units.
- Starting current is to be supplied first by one battery unit and then the other on successive operations of the starter. The changeover is to be made automatically, except for manual start.
- In the event the engine does not start after completion of its "attempt to start" cycle, a visible indicator and audible alarm will sound on the controller.
- In the event that one battery in inoperative or missing, the control is to lock-in on the remaining battery unit during the cranking sequence.

Automatic shutdown following automatic start is accomplished as follows:

- If the controller is set up for automatic engine shutdown, the controller shuts the engine down only after all starting causes have returned to normal and a 30-min minimum run time has elapsed.
- When the engine emergency overspeed device operates, the controller removes power from the engine running devices, prevents further cranking, energizes the overspeed alarm, and locks-out until manually reset. Resetting of the overspeed circuit is required at the engine and resetting the controller main switch to the OFF position.
- The engine will not shut down automatically on high water temperature or low oil pressure when any starting cause exists. If no other staring cause exists during engine test, shutdown is permitted.
- The controller is not capable of being reset until the engine overspeed shutdown device is manually reset.

Acceptance testing, performance, and maintenance

Field acceptance tests. The pump manufacturer, driver manufacturer, controller manufacturer, and transfer switch manufacturer, or their designated representatives must be present for this test, and the authority having jurisdiction must be notified of the time and place of the test.

All electric wiring to the firepump motor(s), jockey pump(s), interwiring, emergency power supply, etc. must be completed and checked by the electrical contractor prior to the initial startup and acceptance test.

Test equipment must be provided to determine the speed, volts, and amps of electric-motor driven pumps. For electric motors operating at rated voltage and frequency, the ampere demand must not exceed the product of the full-load ampere rating times the allowed service factor stamped on the motor nameplate. For electric motors operating under varying voltage, the product of the actual voltage and current demand must not exceed the product of the rated voltage and rated full load current times the allowable service factor. The voltage at the motor must not vary more than 5% below or 10% above rated nameplate voltage during the test.

Firepump controllers are to be tested according to the manufacturer's recommended test procedure. As a minimum, no less than 10 automatic and 10 manual operations are to be performed during the acceptance test. The automatic operation sequence of the controller must start the pump from all provided starting features, including the pressure switches or remote starting signals. Tests of engine-drive controllers are to be divided between both sets of batteries.

The selection, size, and setting of all overcurrent protective devices in the system (including those in the controller, are to be confirmed to be in accordance with the requirements specified in NFPA 20.

The fire pump must be started from each power service and run for a minimum of 5 minutes. Manual emergency operation is to be accomplished by a manual actuation to the fully latched position in a continuous motion. The handle is to be latched for the duration of this test.

On installations with an alternate source of power and an automatic transfer switch, loss of the primary source must be simulated and transfer occur while the pump is operating at peak load. Transfer from normal to emergency source and retransfer must not cause opening of overcurrent protective devices in either line. At least half of the manual and automatic operations are to be performed with the fire pump connected to the alternate source.

If the alternate power source is a generator set, the acceptance test for it must be in accordance with NFPA 110, *Standard for Emergency and Standby Power Systems*.

Fire pumps must be inspected, tested, and maintained in accordance with NFPA 25, *Standard for the Inspection, Testing, and Maintenance of Water-Based Fire Protection Systems.*